# Guarding Your Business

## A Management Approach to Security

# Guarding Your Business

## A Management Approach to Security


Edited by

**Sumit Ghosh**
**Manu Malek**

and

**Edward A. Stohr**

*Stevens Institute of Technology*
*Hoboken, New Jersey*



**Kluwer Academic / Plenum Publishers**
**New York, Boston, Dordrecht, London, Moscow**

Library of Congress Cataloging-in-Publication Data

Management of Technology Symposium (2002: Stevens Institute of Technology)
Guarding your business: a management approach to security/edited by Sumit Ghosh, Manu Malek, Edward A. Stohr.
p. cm.
"Proceedings of the Management of Technology (MOT) Symposium held at Stevens Institute of Technology during October 22–24, 2002"—Introd.
Includes bibliographical references and index.
ISBN 0-306-48494-3
1. Business enterprises—Computer networks—Security measures—Congresses. 2. Electronic commerce—Security measures—Congresses. 3. Computer networks—Security measures—Congresses. 4. Computer security—Congresses. 5. Data protection—Congresses. 6. Business enterprises—Computer networks—Security measures—United States—Congresses. 7. Electronic commerce—Security measures—United States—Congresses. 8. Computer networks—Security measures—United States—Congresses. 9. Computer security—United States—Congresses. 10. Data protection—United States—Congresses. I. Ghosh, Sumit, 1958– II. Malek-Zavarei, Manu. III. Stohr, Edward A., 1936– IV. Title.

HD30.38.M325 2002
658.4'78—dc22

2004044179

ISBN 0-306-48494-3


233 Spring Street, New York, New York 10013

http://www.wkap.nl/

10 9 8 7 6 5 4 3 2 1

A C.I.P. record for this book is available from the Library of Congress



Printed in the United States of America

# PREFACE

## INTRODUCTION

The industrial landscape of yesterday was a collage of companies with diverse products and resources, the only common element between them being people. Today, and increasingly in the future, virtually all organizations will feature computing and intelligent networking infrastructure as a common and one of the most important resource. This is not entirely surprising since computers represent an extension of the human mind while networked systems reflect the essential architecture of human civilization. The result has been mixed. While the industries' reach now extends far and wide, a new kind of threat looms on the horizon. Loosely encompassed by the terms cyber-threat and cyber-crimes, these threats are extremely powerful, occur rapidly, and are often subtle and elusive.

In the last two decades, increasingly banks and financial institutions have been conducting transactions over the Internet using different geographically dispersed computer systems. Today, businesses that accept transactions via the Internet gain a competitive edge by reaching a worldwide audience at relatively low cost. The Internet, however, poses a unique set of security issues that businesses must address at the outset to minimize risk. Customers will submit information via the Web only if they are confident that their personal information, such as credit card numbers, financial data, or medical history, is secure. As we are continuing to learn, security is not always a given and, most important, it is not absolute. For example, in July 2001 Microsoft conceded that all versions of Windows 2000 and early "beta" versions of its new XP operating system contained a "serious vulnerability" that allows hackers to take control of victims' machines. Microsoft promised to cure the problem before XP's rollout. But reportedly, after a few hours of the announcement of Windows XP on October 10, 2001, it was hacked! The threats, risks, and failures in the computing and networking world mirror those in the physical world, but the risk factors in the electronic world are different. Failures are more widespread; failures stemming from attacks are harder to predict; and the effects of attacks are devastating. Paper data, even where public, is hard to search and correlate. In contrast, electronic data can be searched easily and networked data can be searched remotely and correlated with other databases.

A 1996 GAO report estimated that 250,000 attacks per year had been perpetrated against the Department of Defense computer systems, and that the rate of attack was increasing by about 100% per year. According to the FedCIRC, the incident handling entity for the federal government, 130,000 government sites totaling more than one million hosts were attacked in 1998. Also, a 2002 survey conducted by the Computer Security Institute and the FBI revealed that 90% of the 507 participating organizations detected computer security breaches within the past 12 months, 74% cited their Internet connections as a frequent point of attack, but only 34% reported the intrusions to law enforcement. These numbers must be taken with a big grain of salt since they only relate to known attacks and vulnerabilities.

A key to preventing security attacks is to understand and identify the IT infrastructure vulnerabilities, and to take corrective action. Where corrective action is difficult, attacks must be monitored to estimate the time and extent of penetration by perpetrators. Also the value of information to an organization must be estimated. Businesses create and collect a tremendous amount of information, such as employee information, customer information, product information, and process information. Such information is widely distributed due to increasing employee mobility, companies migrating from a centralized to a distributed computing environment, companies growing their reach geographically, and as a result of employee mobility. Information is a major business asset: the nature and quality of information, and how effectively it is used, distinguishes an organization from its competitor.

## OBJECTIVE OF THE BOOK

This book encapsulates the proceedings of the Management of Technology (MOT) Symposium held at Stevens Institute of Technology during October 22–24, 2002 and is the first in a planned series of books that will accompany Stevens' MOT Symposium Series (http://attila.stevens-tech.edu/motsymposium).

The title of the book reflects the symposium's focus in 2002, namely, security (the issues that managers must address to safeguard their organizations against potentially devastating electronic attacks) architecture for business. Subsequent symposia will cover other topics within the broad framework of technology management.

The objective of this book is to provide business managers a complete and comprehensive awareness of basic network security issues at the correct level of detail so they can make decisions that are sound from both scientific and business perspectives. Subsequent symposia will cover other topics within the broad framework of technology management, grounded in science and engineering. Selected symposium speakers and other experts were invited to write the chapters for this volume in their specific areas of expertise and were requested to add case studies to strengthen coverage of the issues.

## OVERVIEW

This book outlines the organizational elements that must be in place to protect the information assets of the organization. These elements include technologies such as encryption, firewalls, biometric devices, and back-up and recovery mechanisms. All of these technical elements must be employed within a comprehensive framework of architecture so as to systematically guard against hostile intrusion, resist human error, and survive physical attack and natural disaster. Attributes of the architectural framework include risk assessment, determination of quality-of-service levels that balance safety versus cost, disaster recovery procedures, determination of access rights to data and software, and a security-conscious culture in the organization. Thus, the main message of the book is that the architecture must parallel the physical structure of the organization that has long been designed to systematically protect the physical assets of the organization. Questions answered by the book include: How can one organize for security? What are the fundamental issues in network security? What organizational structures, policies, and procedures must be in place? What legal and privacy issues must be addressed?

## ORGANIZATION OF THE BOOK

The book is divided into four parts, Part I through Part IV. Part 1 presents background information through two chapters. Chapter 1 includes an overview of the security threats and techniques and explains authentication, encryption, and public-private key mechanisms in detail. Chapter 2 examines the nature of network security from a fundamental perspective and develops a systematic logical analysis to present a clear picture of the past and a glimpse of the future.

Part II addresses security threats to organizations in three chapters. Chapter 3 enumerates the actual experiences of the Commanding Officer of the Computer Crimes Investigation Unit of the New York Police. This eye-opening chapter reveals how quickly most passwords may be cracked automatically, even those borrowed from dictionaries in foreign languages, and how perpetrators may leave silent background scripts running on computers to capture every keystroke of an unsuspecting user. Chapter 4 describes the results of a war-gaming exercise involving cyber-attacks on the nation's networking infrastructure including telecommunications, Internet, electric grid, and financial services. Chapter 5 explains U.S. laws that are designed to fight security threats and expresses justifiable concern that laws recently passed by U.S. Congress in the wake of the 9/11 incident may seriously undermine U.S. Citizens' right to privacy as guaranteed by the U.S. Constitution and the Bill of Rights.

Part III reports on the state-of-the-art in security technologies and includes three chapters. Chapter 6 focuses on the specific security needs of the financial industry including protecting information, defeating denial of service, and preventing fraud and identity theft. Chapter 7 reviews the current state-of-the-art in wireless wide-area, local-area, and personal-area networks. Chapter 8 explains the difficulties and tradeoffs underlying the use of passwords, biometrics, and other methods of authenticating human users.

Part IV describes new methodologies to organize and manage security and comprises three chapters. Chapter 9 discusses the different dimensions of risk management while Chapter 10 is critical of the current security policy that is administered on a piece-meal ad hoc basis. Consequently, different components of network security do not work with one another. The chapter describes the experience with developing and evaluating a smart firewall project. Chapter 11 describes actual experiences related to identifying and closing key security gaps in the Internal Revenue Service's operations. Finally, Chapter 12 reviews in detail the different processes and procedures needed to maintain security in an organization. In essence, it describes the organizational architecture that needs to be in place to guard your business against the many security threats outlined in earlier chapters of the book.

Sumit Ghosh, Manu Malek, and Edward A. Stohr
*Stevens Institute of Technology*

# ACKNOWLEDGMENTS

The editors express their sincere gratitude to all the speakers, contributing authors, and attendees of the 2002 MOT symposium, whose enthusiasm made the symposium a success. Special thanks are due to the organizing committee at Stevens Institute of Technology including Chris Bullen, John Byrne, Stan Clark, Audrey Curtis, Hosein Fallah, Sharen Glennon, Jerry Hultin, John Keating, Rajkumar Kempaiah, Louis Laucirica, B.J. Taylor, Melissa Vinch, and Elizabeth Watson. External advisory board members who deserve our thanks are: Ed Amoroso (AT&T), Ed Cannon (Grey Global), David Luce (Rockefeller Group Inc.) and Daniel Schutzer (Citigroup). Ramaswamy Iyer provided invaluable services in editing and compiling the book. Ana Bozicevic of the editorial department and the production staff at Kluwer Academic/Plenum Publishers deserves our sincerest appreciation.

# CONTENTS

Chapter 1

# SECURITY PRINCIPLES FOR MANAGERS

Manu Malek
*Department of Computer Science*
*Stevens Institute of Technology*

**Abstract:** This chapter addresses the importance of information security, provides some statistics about security breaches and threats, and discusses security services and mechanisms. One of the important security mechanisms is, of course, encryption. An overview of secret key and public key encryption techniques is provided, followed by a brief description of other security services such as authentication, message integrity, and non-repudiation.

**Key words:** security, information security, security services, security mechanisms

## 1. INTRODUCTION

Organizations and enterprises collect many different types of information, such as employee information, product information, and shareholder information. This information is a major business asset, and must be safeguarded since effective use of information provides a competitive advantage.

In e-business and transactions on the Internet, buyers and sellers are both very concerned with security. Buyers want to be assured that the private information they put on the net, such as their credit card numbers, is secure. Sellers are concerned that the data in their back-office databases, processes, and applications is not corrupted or stolen. A major security issue is that the Internet is an open network, and this openness works against security: the more open an environment, the more security problems. The Internet is distributed, open, and public; desirable properties that also create many security problems.

Figure 1 shows some statistics about the security breaches in different areas. The data summarizes the result of the FBI/Computer Security Institute Year 2001 questionnaire sent to about 500 respondents. For example, it shows that 93 percent of the respondents reported that they were attacked by viruses in 2001, and this reflected an increase of 35% with respect to Year 2000. As shown, the largest increase was in financial fraud, mostly attributable to on-line auctions.

Figure 2 provides another look at security intrusions and vulnerabilities. Intrusions were relatively few in the early1990s, but there has been a major increase since 2000. About 25,000 intrusions were reported in the Year 2000 (keep in mind that not all enterprises that suffer security breaches report them). The line moving upward in this figure shows various types of threats, starting with very simple ones in the early '90s, like password guessing. The sophistication of attacks increased with self-replicating codes, such as viruses, then password cracking (where the cryptographic password is cracked), and on to the more sophisticated threats shown in Figure 2. In February 2000, major distributed denial-of-service attacks were perpetrated on some high-profile web sites including Yahoo.com, Amazon.com, and Ebay.com. More recently, there have been attacks by Internet worms like "Code Red". Against this rising sophistication in threats, we have a lot of hacking tools that are available: hackers no longer have to be experts in computer science or security; they could use available tools. For example, a tool such as *nmap* [2] can be used to find all the open ports, a first stage in an attack. So the level of knowledge that is required of the attackers is coming down, while the sophistication of the attacks is going up, giving rise to major security concerns.

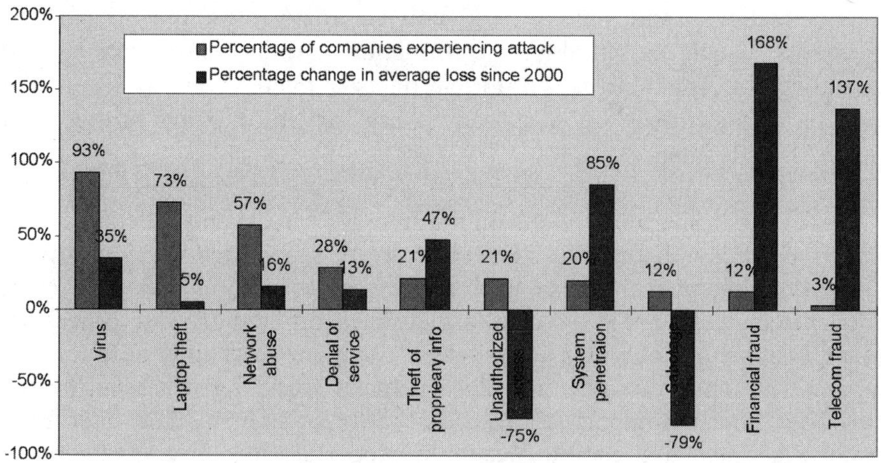


Figure 1. Security breaches.

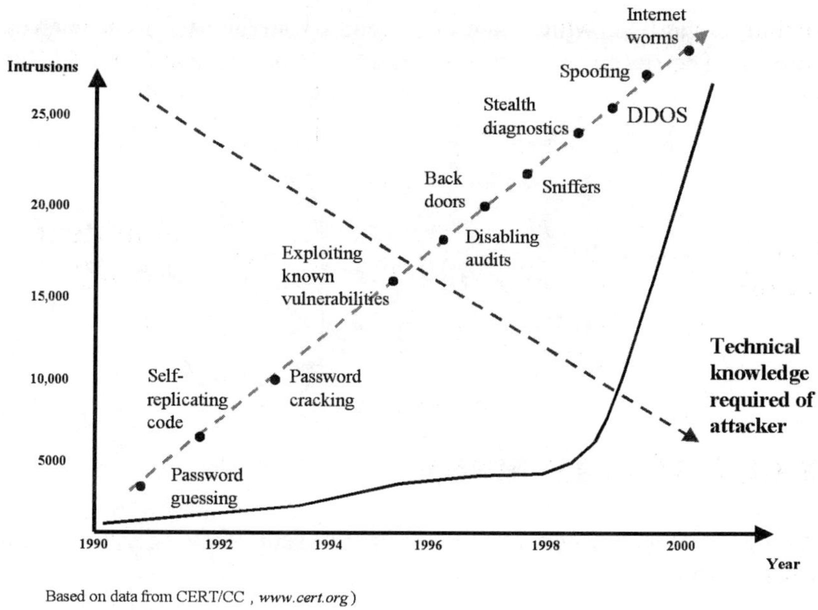


Figure 2. Security vulnerabilities and threats (based on data from CERT/CC [1]).

The remainder of this chapter describes security services and mechanisms. It starts with encryption, and provides an overview of secret key and public key encryption techniques. It then continues with a brief description of other security services, namely, authentication, message integrity, and non-repudiation, and the corresponding mechanisms.

There are four security services [3]:

*Confidentiality*, which has to do with keeping a message secret so that no one can access the message except the sender and the intended receiver

*Authentication*, to make sure that the entities, whether human users or software, are indeed who they claim to be

*Integrity*, which has to do with the integrity and authenticity of the message; to make sure that the message has not been tampered with during transmission

*Non-repudiation* or *non-denial*, to make sure that the sender of a message cannot deny sending it, or the receiver of a message cannot deny receiving it

Security mechanisms are used to implement security services. There is no single mechanism that can provide all of the security services. Encryption, of course, provides confidentiality: if a message (referred to as plaintext or cleartext) is encrypted using some key, then the message is converted into a corresponding ciphertext, which cannot be read by anyone without that key (see Figure 3). The strength of encryption depends on the size of the key as we'll see.

Figure 3. Encryption.

# 2. ENCRYPTION TECHNIQUES

The two general techniques for encryption are secret-key encryption and public-key encryption.

## 2.1 Secret-Key Encryption

In secret-key encryption, also referred to as symmetric encryption, the same key is used by both the sender and the receiver. The two sides must pre-coordinate to obtain the secret key that they are going to use; ideally no one else has access to that key. In fact, anyone else with the key can read that message. Figure 4 shows a sender sending a plaintext message P, which is encrypted using some encryption algorithm with a key K. The resulting cipher text C is transmitted to the receiver. The receiver uses the same encryption algorithm and key to decrypt and recover the plaintext. Since both sides use the same key, the method is called *symmetric*. Secret-key encryption has been in use for a long time, in fact since the times of the Roman Empire.

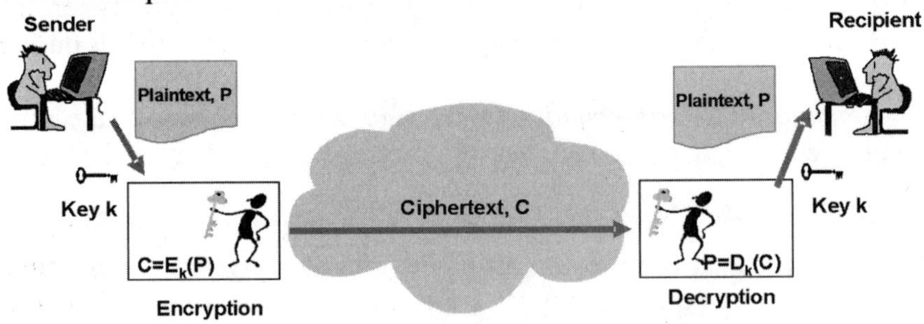


Figure 4. Secret-key encryption.

There are several examples of secret-key encryption, the most popular being Data Encryption Standard (DES). Other examples are RC2, RC4, and RC5 [4].

All secret key encryption systems are based on substitution and transposition. In substitution, each element of plaintext is mapped into another element. For example, if one thinks of plaintext having characters A, B, C, D, then one can substitute, for example, B for A, C for B, and so forth, i.e., shift characters by one (the key). In transposition, the plaintext characters are first organized into groups of n characters each; then within each group the characters are switched around according to some rule. For example, a group including ABC may be changed to CBA, and so forth. Then the number of characters in each group (n) and the rule for transposing them will be the key.

Secret-key encryption systems, including DES, use many stages of substitution and transposition. Of course, the examples given above were in terms of characters, but in today's secret-key encryption systems, we deal with bits (you can think of characters having been encoded in ASCII, for example).

DES is a very popular symmetric encryption algorithm. It was initiated by NBS (National Bureau of Standards, now called NIST, National Institute of Standards and Technology) in 1977, then updated and standardized for non-military applications in 1993. DES is a block cipher; that is, the plaintext is divided into blocks of 64 bits and every block goes through the DES process to produce a 64-bit ciphertext. So the plaintext blocks go in sequentially and the corresponding ciphertext blocks come out sequentially. If a 64-bit plaintext block is applied to a DES stage with key K, and then the resulting 64-bit ciphertext is applied back into the same DES stage, one would get the plaintext back (see Figure 5) [see [4] for details].

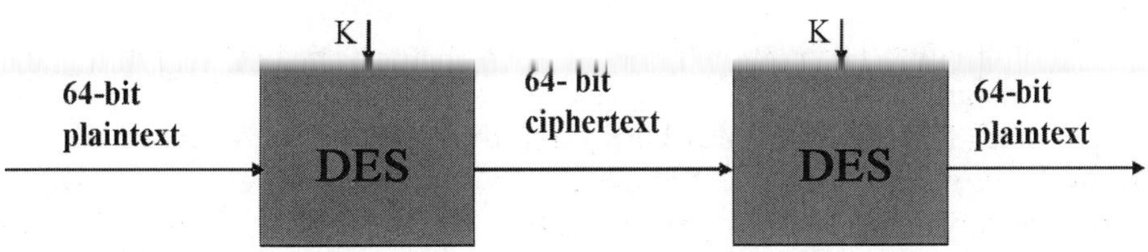


Figure 5. Data Encryption Standard (DES).

The initial DES used a 56-bit key, which is not very secure any more. In fact, there is a controversy about this: DES is based on IBM's Lucifer, which used a 128-bit key. But to standardize it, the government did not want the key to be 128 bits and insisted on a 56-bit key (actually a 64-bit key, but one out of every 8 bits is used for parity). So DES ended up as a 56-bit key system. The government's argument was that they wanted to be able to decrypt any message (e.g., for law enforcement or national security reasons) within reasonable time and available resources.

At the time that DES was standardized in 1977, it was a very secure algorithm. With the improvements in VLSI technology over the years, now the 56-bit key DES can be broken by brute force (i.e. by trying all the combinations of the 56-bit key) in about 10 hours [see Table 1]. In fact, a DES-cracking challenge was announced in 1997 by RSA Security with a $10,000 dollar reward. It took a group of scientists 96 days to crack it. A similar challenge was repeated in 1999, and it was done in twenty-two hours [5].

Table1. Strength of DES [4]

| Key Size (bits) | Number of alternative keys | Time required at 1 decryption/μs |
|---|---|---|
| 32 | $2^{32} = 4.3 \times 10^9$ | 2.15 milliseconds |
| 56 | $2^{56} = 7.2 \times 10^{16}$ | 10 hours |
| 128 | $2^{128} = 3.4 \times 10^{38}$ | $5.4 \times 10^{18}$ years |
| 168 | $2^{168} = 3.7 \times 10^{50}$ | $5.9 \times 10^{30}$ years |

To strengthen DES, some changes were made to it. One of the changes was to design triple DES (3DES), referred to also as Triple Data Encryption Algorithm (TDEA). It applies DES three times with three different keys as shown in Figure 6. The effect is essentially an encryption algorithm with an effective key size 3 times that of DES, or 168 bits. Triple DES was standardized by FIPS with an effective key size of 2X56 = 112 bits by setting K1 and K3 equal in 3DES (see Figure 6). Also NIST standardized triple DES for PPP (Point to Point Protocol, the protocol using TCP/IP over a dial-up line). The decryption process is the reverse, as shown in Figure 6.

Note that a reverse stage is used in the middle; in other words, the direction of the middle stage is changed (that is why it is shown as D in Figure 6). Then of course the reverse occurs in the decryption process.

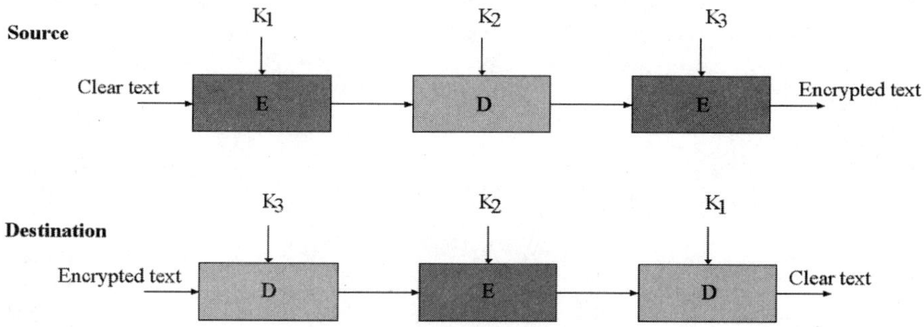


Figure 6. Triple DES.

Another way to improve DES is Cipher Block Chaining (CBC). DES, being a block cipher, creates sequential 64-bit ciphertext blocks from plaintext blocks. It is potentially possible to switch the ciphertext blocks and deceive a receiver. In other words, one could attack the integrity of the message even if it may show at the receiving end that the confidentiality is satisfied. Cipher block chaining mixes up ciphertext blocks so that the tight correspondence between the blocks is broken [see Ref. 4 for details].

## 2.2. Public-key encryption

In public-key encryption there are two keys, a public key and a private key. If a user wants to send an encrypted message, the plaintext is encrypted with the public key of the receiver to create the ciphertext. The ciphertext is then decrypted at the receiving end using the private key of the receiver (see Figure 7). This method of encryption is also referred to as asymmetric encryption because different keys are used to encrypt and decrypt. Note that only the receiver who has that private key can decrypt the ciphertext. A simple analog to this in the physical world is as follows. If you think of a combination lock, anyone can lock it, but only the person who knows the combination can unlock it. (Using a similar analogy for secret-key, encryption is a lock and key where the key is used to lock as well as unlock.)

Public key encryption is a very clever scheme that was invented by Diffie and Hellman in the mid 70's. They initially came up with the idea and put down the requirements for public key encryption without having any algorithm for it. There are now different algorithms to implement public key encryption, a popular one being RSA (RSA is the acronym created by the last names Rivest, Shamir, and Adelman).

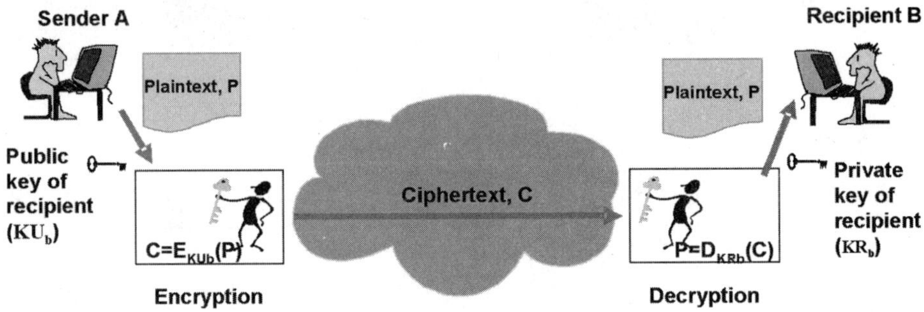


Figure 7. Public key (or asymmetric) encryption.

## Differences with Secret-Key Encryption

The advantage of public key encryption is that, unlike in secret-key encryption, there is no need to pre-coordinate a key between the two sides. The public key can be publicized, and anyone who wants to send an encrypted message can use that public key to encrypt it. Only the entity with the private key can decrypt it. So unlike secret key systems where this pre-coordination is needed and the security of the key is essential, here the key for encryption is public. Therefore, people can exchange secure communication even if they have not met, or do not even know each other. If one wants to communicate with a bank, for example, as long as one knows the public key of the bank, one can send secure messages to the bank.

Unlike secret-key systems where the algorithms are based on simple substitution and transposition, in public-key encryption, the algorithms have deep mathematical bases; they are based on properties of prime numbers, modulus algebra, and exponents. Yet another difference with secret-key encryption is that public keys are much longer: while secret keys are in the order of 128 bits, public keys are 512 bits or longer.

# 3. DISTRIBUTION OF PUBLIC KEYS

Public keys can be broadcast to the public at large. For example, a bank may advertise a public key, and anyone who wants to send a secure message to the bank can use the public key and encrypt the message. But how can one be sure that the public key really belongs to that bank?

Digital certificates are used for this purpose. A digital certificate (or a public key certificate) is a certificate issued by a trusted third party that certifies the ownership of the public key. Of course, how much one can trust in that digital certificate depends on the reputation of the entity (the Certification Authority, CA) that issued the certificate. It is analogous to a notary public certifying that a signature belongs to a person by signing the certificate. Similarly, a certification authority issues the certificate, and digitally signs it (digital signature will be discussed shortly). Versign and E-Trust are examples of certification authorities.

An individual or a company can apply for a digital certificate by providing their public key and their identifying information to a certification authority. The certification authority creates a certificate with the name of the entity, the public key of the entity, a serial number, and the expiration date for the key. The certificate is digitally signed by the certification authority (see Figure 8). The digital signature (to be discussed in Section 6) allows for verification of the validity of the certificate (that of course, depending on the reputation of the CA). Some web sites display a small icon indicating "certificate," which when clicked on would display the contents of the certificate.

Figure 8. Digital certificate.

## 4. USER AUTHENTICATION

Authentication is the process of validating the identity of a user or a process. It applies to both human users and machines. Here we focus on user authentication. An individual can be authenticated by something that the individual has, like an identification card, some characteristic of the user, like a fingerprint, or something the user knows, like a password.

Passwords require a previously communicated shared secret. There are different types of password. A reusable password is used for each authentication. It should be easy to remember, but difficult to guess. System administrators make users change their reusable passwords periodically to prevent compromise.

A one-time password includes two parts: one part is a randomly generated number, and the other part is a set of characters that can be easily remembered by the user. The so-called secure tokens are used for this purpose and provided to the users. Secure tokens are devices with a small IC chip capable of generating a random-number that changes every 20 seconds or so. That random number must of course be synchronized with that generated by the server. To get authenticated, the user enters the easily-remembered part followed by the random number indicated on the secure token.

## 5. MESSAGE INTEGRITY

Message integrity involves making sure that a message has not been tampered with during transmission. One might think that using encryption (which provides confidentiality) would guarantee message integrity, but it doesn't. For example, in DES, a block encryption scheme, it is possible to switch the ciphertext blocks but the decrypted ciphertext may still make sense to the receiver. That would mean the message has been modified (the integrity of the message has been attacked), but still the receiving end is satisfied of its confidentiality.

Message integrity mechanisms use a small amount of data as the "fingerprint" or "digest" of the message. The message is sent followed by this fingerprint to the receiver. At the receiving end the receiver uses the same operation on the message to calculate the fingerprint of the message, and compares it with the received fingerprint. If the two fingerprints agree, then the receiving end can be reasonably sure of message integrity.

An example of an algorithm to generate the digest of a message is Message Digest 5 (MD5). It creates a 128-bit digest (or hash) of any length message. To start with, the message has to be padded if necessary to make it a multiple of 512 bits. As shown in Figure 9, the digest calculation starts with the digest value initialized to 128 bits (e.g., 128 zeros). This value is combined with the first 512 bits of the message, using a complex transformation, to produce a new 128-bit value for the digest, and so on. For example, one could do a bit-by-bit exclusive-or of each bit of the initial digest with every 4th bit of the first 512-bit to get the next 128-bit digest, and so on.

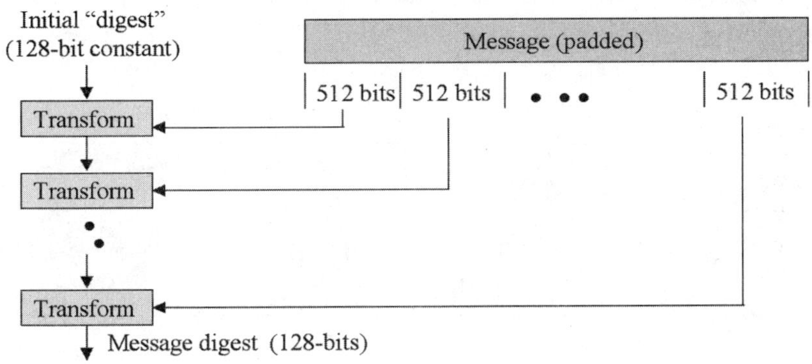


Figure 9. The MD5 Process.

## 6. NON-REPUDIATION OR NON-DENIAL

Now non-denial is a legal term, and it has several aspects to it. The basis for it, however, is digital signature. A method for digitally signing a message based on public key encryption will be described briefly next.

Recall that in public-key encryption, one encrypts a message with the public key of the receiver, and the receiving entity decrypts it with its private key. To support digital signature, there is an additional requirement on the public-key system: it should be possible to encrypt a message with a private key and decrypt with the corresponding public key. Thus, to send a digitally signed message, the sender would encrypt the message with the sender's private key (i.e., digitally sign it); then at the receiving end the receiver can decrypt it with the sender's public key. Note that only the sender could have thus encrypted (signed) the message, because only the sender has access to the sender's private key.

Figure 10 shows the process of sending a digitally signed encrypted message by a sender, and shows how the receiver would extract the plaintext back. The sender first digitally signs the plaintext (encrypts the plaintext with the sender's private key), then encrypts the signed message with the receiver's public key. This ciphertext is transferred to the receiver, who would first use the receiver's private key to extract the signed message, and then would decrypt this signed message with the sender's public key to extract the plaintext. Thus the receiver would have both the plaintext and the signed plaintext. Using them the receiver can potentially prove that the sender indeed sent that plaintext message. That is the basis for non-denial. It must be combined with time-stamp and sequence number to protect against replay attack (i.e., an attacker capturing this scenario and replaying it later).

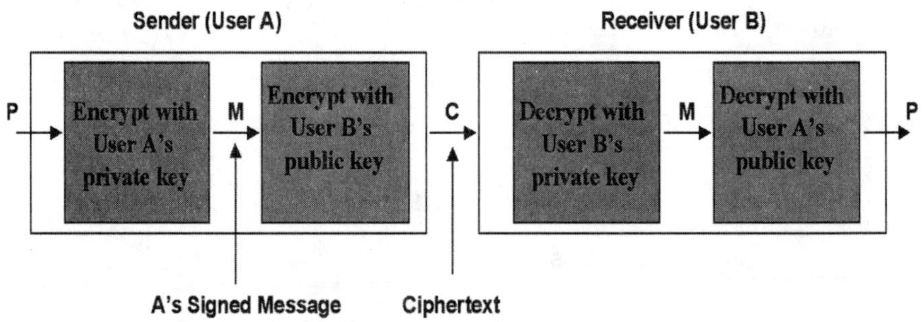


Figure 10. Digital signature: basis for non-denial.

## 7. SUMMARY

This chapter has provided an introductory tutorial on the importance of security, security services and mechanisms. The chapter does not cover network or application security concepts and methods (such as IPSec, Secure Socket Layer, SSL, and Secure Electronic Transaction, SET). Such methods may be used in typical transaction on the Web, such as a simple credit-card purchase, or business-to-business financial transactions. The following chapters cover some related concepts and techniques.

## REFERENCES

1. Computer Emergency Response Team/Coordination Center (CERT/CC) at Carnegie Mellon University's Software Engineering Institute, http://www.cert.org/

2. www.insecure.org/nmap/nmap-fingerprinting-article.html
3. www.trusecure.com
4. W. Stallings, Network Security Essentials, (2nd ed.), Prentice Hall, 2003.
5. Search for DES cracker on www.eff.org

## About the Author:

Manu Malek is Industry Professor of Computer Science and Director of the Certificate in CyberSecurity Program at Stevens Institute of Technology. Prior to joining Stevens he was a Distinguished Member of the Technical Staff at Lucent Technologies Bell Laboratories. He has held various academic positions in the US and overseas, as well as technical management positions with Bellcore (now Telcordia Technologies) and AT&T Bell Laboratories. He is a fellow of the IEEE, an IEEE Communications Society Distinguished Lecturer, and the founder and Editor-in-Chief of the *Journal of Network and Systems Management.* He earned his Ph.D. in EE/CS from University of California, Berkeley.

Chapter 2

# THE FUNDAMENTAL NATURE OF NETWORK SECURITY

Sumit Ghosh, Hattrick Chair Professor
*Department of Electrical & Computer Engineering*
*Stevens Institute of Technology*

**Abstract**: This is one of the tutorial presentations on the first day of the MOT symposium. It is mostly verbatim, with some portions edited. It analyzes network security from a fundamental perspective and traces the evolution of the key issues.

**Keywords**: Network security, information security, computer viruses, networking, and computational intelligence.

## 1. INTRODUCTION

In this chapter, I examine network security principles from a fundamental point of view by critically examining the basic elements of any network: its nodes, its links, and the control algorithm that moves information around the network. Viewing these network elements from a fresh perspective yields new insights and lessons for those of us interested in comprehensive security. This examination suggests directions for future secure network design.

## 2. MY RESEARCH TRAINING

I started off at the graduate level working in the discipline of hardware description languages (HDLs), which deals with how to write languages to design computers and digital hardware. That was the subject of my Ph.D. thesis at Stanford many years ago. Thereafter, I became very interested in how hardware works and how HDLs should really work, but we did not have access to a parallel processor at Stanford at that time. Subsequently, when I joined Bell Labs Research, I had access to parallel processors, started working with them, and did some very interesting research in the area of modeling and simulation. Then I moved to Brown University and at that time Bellcore, now Telcordia, approached me, stating that they understood quite well how a single switch of a network behaves, but not the behavior of multiple switches in a network, the causes of congestion, what congestion really means, and so forth. They asked me to look at modeling and simulation of a large-scale broadband-ISDN network, presently referred to as an ATM network. So we became engaged in that effort for about three years and the research was done primarily with undergraduate students at Brown. The Army Research Office and the Ballistic Missile Defense Organization supported us and were very pleased with the results. They stated that our effort provided a model, namely large-scale modeling and asynchronous distributed simulation, for the US Department of Defense to conduct its future research in complex systems.

## 3. A UNIQUE PERSPECTIVE INTO NETWORK SECURITY

My next move was to Arizona State University where I met a Ph.D. student with a very unusual background. He was a major in the US Army and had won a fellowship to pursue his Ph.D. and then join the US Military Academy at West Point as a professor. He had been the chief of networking at the White House during President Bill Clinton's first term and had sole responsibility of developing and maintaining a secure network, ready to be deployed anywhere outside the US so that the president could remain in constant touch with the White House, National Security Council, and the Pentagon, during a foreign trip. That was his job. He had access to state of the art equipment, which was asynchronous transport mode (ATM) networking that, at the time, had an extremely high degree of sophisticated encryption. This was several years ago. Although he deployed the best technology money could buy, he felt there was a serious problem. He would tell me that he had all these encryption algorithms, million dollar devices, but what would prevent someone from simply snapping the wire? With the wire snapped, there is no longer a connection, and the president cannot talk to his national security advisers. Even if the message is highly encrypted, it just does not make any sense. There are so many ways in which a communication can be interrupted. Encryption is surely one of those ways but not the only way and sometimes offers us a false sense of confidence. There should be other ways that can we come up with, which will render networks more secure.

## 4. WORKING WITH THE NATIONAL SECURITY AGENCY

So, with that, we started to work on how to make networks secure, or harden them. At that very moment, the National Security Agency invited a group of approximately 150 people, including myself from ASU, to attend a three-day workshop to review the network rating model (NRM) effort. NRM, a framework aimed at securing a network, had been developed by a group of 12 individuals at NSA over the past two years. NSA's objective was for NRM to be embraced by industry, military, government, and educational institutions. As the NSA team started to present their idea, it became quite clear after about fifteen minutes that NRM was severely limited in its vision and scope. The issue was serious and realizing the gravity of the situation, the principal organizer canceled the remainder of the planned agenda, organized the workshop attendees into groups, and charged each group with the task of solving the problem in an innovative manner. The next few days were filled with intense and engaging discussions. Following our return to ASU, we started to work on the problem and, quite by accident, we hit upon a very promising idea -- a fundamental security framework [2, 3]. Over the next two years, the NSA invited us twice in much smaller, authors' group meetings, eventually resulting in NSA adopting the idea of the fundamental security framework.

## 5. A COMPREHENSIVE VIEW OF SECURITY

Security is a very comprehensive concept. Looking back, I think the reason we were able to come up with the framework can be summarized as follows. My Ph.D. student's actual experience with state-of-the-art secure networks and his insights about security combined with my broad research experience in hardware description languages, networking, modeling and simulation, large-scale software, control and coordination algorithms for complex systems, intelligent transportation systems, and biological systems, generated in us a unique insight into the comprehensive nature of network security. I will now present a few hypothetical scenarios to help explain this point.

Assume that I am engaged in an on-line purchase of an item using my credit card. I access a secure Internet site by typing in the URL address beginning with ``https://", then type the credit card number, etc. to complete the transaction. I would hope, understandably, that all will go well with my transaction and that no one else will gain access to my credit card number. This is exactly what the on-line store and my credit-card company wants me to believe. I do believe them. However, entertain the following scenario.

In any IP network, which is the basis of the Internet, there is no such thing as a direct connection between myself and the on-line store. One has to connect through many intermediate nodes or routers, which participate in routing the IP packets constituting the transaction, to the destination. Now, according to a fundamental principle, every one of the intermediate nodes must intercept my packets for processing. So let's say that I am trying to buy something from a vendor's secure website, yet one of the nodes through which my IP packets pass is located in New York City and there's a perpetrator sitting there. The perpetrator does not do me any harm except simply storing the encrypted contents of the IP packets on a hard drive. The activity appears harmless, even silly, since the packets are encrypted and the perpetrator cannot decipher the credit card numbers. The perpetrator continues this activity and over a period of say two years and accumulates over two million credit card transactions. By this time, someone may come up with a clever technique to break the encryption algorithm. This is entirely conceivable, because of fundamental weaknesses with encryption algorithms. First, there are no mathematical proofs underlying the infallibility of these algorithms. Consequently, clever individuals with good insights often crack these algorithms in a matter of hours to days. Second, the strength of encryption algorithms is derived from computational intractability and this is clearly threatened by advances in computing engines. Now, the perpetrator has two million credit card numbers and other critical details in his/her possession. For the sake of argument, assume that half of these cards have been cancelled or expired, leaving behind a million valid credit card numbers. If a coordinated attack is now launched on these million credit card numbers, distributed throughout the world, the effect will be catastrophic. This can occur today, despite the best encryption, and very few, if any, banking systems could withstand the onslaught.

# 6. A FUNDAMENTAL UNDERSTANDING OF NETWORKING

So the point is, security is not just about any one view, but far more encompassing, a very broad perspective, and that's the message I wish to convey in this paper. Without the broad perspective, things can go very wrong. I will make an effort to present to you a fundamental understanding of the nature of networking [6, 7] and examine logically where it is headed in the future. Network security is not a mere fad; it is not something that is here today, gone tomorrow. It is very fundamental. Naturally, the question is why? To understand the answer, consider a networked system, or any system, given that increasingly nearly every system is evolving towards a networked system, and ask the following. What does a networked system consist of, fundamentally? First, it has nodes that are units basically responsible for all the processing. Examples include IP and ATM switches developed by vendors such as Lucent, Cisco, Juniper systems, Fore/Marconi, etc. Second, a networked system contains links, the basic transmission media, realized in the form of copper cables, optical fibers, sight-to-sight laser transmissions, satellite channels, or GHz wireless links. Third, there are control algorithms that constitute the heart of the system and make it work. To get a feeling for the complexity of the control algorithms, observe that the Internet is distributed all over the world. Parts of the Internet are in the U.S., other parts are in Africa, Asia, Russia, Europe, etc. How do they communicate with each other? It is obvious that the control algorithms play a key role yet, according to many experts, we don't completely understand how they work. If you find this confusing, please keep in mind that while we are completely dependent on the power grid, we still don't know in what sequence to bring online the power generating stations around the country, following a complete shutdown or failure of the grid, such that the stations will not mutually trip each other. Returning back to the problem at hand, clearly, our ignorance of the control algorithms is partly to blame for so many attacks on the Internet.

## 6.1. First Key Element of Networking--Node

Next we focus on the nodes. These are basically computing entities - all they do is compute numbers. Fundamentally, however, these nodes are performing the switching function. To understand the nature of the nodes, let us travel back to the 1920s when the telephone system was the technology showpiece. Back then, the telephone system was kept afloat by human operators. Many readers may recall old black and white TV shows such as "The Untouchables" where Elliot Ness, a federal agent is attempting to make a phone call to Florida. These old films show how the operators connect and disconnect specific cables to a box to establish the connection between Chicago and Florida. The human operators were supplying something critical -- intelligence. Today, we have replaced that biology-based intelligence with computers supplying computational intelligence. This intelligence, I will submit to you, represents a fundamental network characteristic. Let me explain why.

## Networks are fundamentally non-scaleable

We will digress just a bit by contrasting two networks. The first is a network that transports packets in pure electromagnetic (EM) form. We will term this an EM network. Second, retail networks owned by say, Macy's or Wal-Mart, where consumer goods are transported between the manufacturers, warehouses, and retail stores. The warehouses stock material items and transport them to appropriate retail stores as they place orders. How do these two networks differ, fundamentally? For the second network, we could place an individual computer, networked with other appropriate computers, on each truck or even each merchandise item that is being shipped from the warehouses to the retail stores or even among retail store.

Consider the first option, namely we place computers on each truck. The computer can determine, dynamically, the best route that the truck may adopt to get to each store such that some important criteria are optimized. The dynamic recomputation of the truck route becomes important when the demand for specific items at key retail stores change on a daily or even hourly basis. To summarize, the intelligence required to determine what optimal route the truck should adopt was provided by the computer sitting in the truck. So if Macy's purchases another company, say, and doubles in size, it is logical to expect the flow of goods to double, requiring the number of trucks to also double. However, along with the increase in the number of delivery trucks, the number of computers also doubles and, thus, the impact on the efficiency of the route computation is minimal. Today, computers are inexpensive, so this scheme is indeed scalable. Beautiful, right?

Now let us examine the EM networks. We transport packets that are not pieces of matter but a quantum piece of energy. To date, however, we have not figured out how to put a computer on top of a packet. In building computers, we have had to deal with silicon, germanium, gallium arsenide, etc. and, someday in the future, we hope to develop computers using optical ingredients or even chemicals and biological material. All of these, however, are matter, and we do not yet know how to place matter on top of a quantum of energy, in transit. Thus, the question remains, where can we perform the computation necessary for every packet? The only answer is at the nodes. Since the number of nodes is finite and much much smaller than the number of packets, fundamentally, an EM network cannot be scalable. How does this relate to security? Well, somewhere one has to verify that the system is working properly. One has to execute code to realize this verification. For example, checking a fingerprint requires computation. The only logical source of computing power is again the node. With all these demands, the nodes often become overloaded.

In this context, let me share a humorous yet real anecdote. I have often sent e-mails to Frank Fernandez at the Howe center, and I have been told that it usually takes three hours. We were joking around one day and suddenly realized that I could simply walk from my building to Howe center is 5 minutes and convey the message faster than email. Under these circumstances, there seems to be little reason to want a network at all. The actual reason for the slow transport is possibly the use of firewalls and other security measures. So, where is the network headed and what is the most likely evolutionary path? I will address this at a later point in my presentation.

## 6.2. Second Key Element of Networking--Link

Our next goal is to secure the links. Once a packet leaves the source, it propagates through optical fiber, copper cables, wireless, etc. The source no longer exercises complete control over the transport. The recipient is generally geographically distant. Here also, the control algorithms play a key role. We will talk about them more, later.

## 6.3. Vulnerabilities of the Nodes

We touched on nodes earlier, let us now focus on their vulnerabilities. If the nodes are already under stress, as we saw earlier and if a perpetrator is aware, what sort of attack can he/she conceive? A simple attack would be to flood the network with bogus packets and stress the system to the point of failure. The nodes do not know a priori which are good and bad packets. A node gets overwhelmed because it attempts to process all the packets. This is a performance attack, now commonly known as a denial of service attack. Also, the nodes are nothing but computers, whether we call them switches or routers. So, we should ask, how are computers vulnerable? Let me a share an anecdote, centered on a question raised by the assistant attorney from the Department of Justice. He told me that he had been talking to a lot of technical people from industry and research labs. No one could explain to him why log files can be deleted. A perpetrator can force into a machine, cause a lot of damage, and then delete the log file, which contained a clear trace of all activities. Thereafter, it would appear as if nothing abnormal had happened. Incidentally, such attacks are common and one such was leveled last week (2002) against our own ECE departmental machine -- Koala. The question is simple, yet very profound. I did some searching and a lot of thinking. It turns out, that starting with the Multics operating system (General Electric and MIT) in the late 1970s and then its successor, Unix (Bell Labs), the underlying thinking was that every piece of information should be organized as a file. This became the core architecture of operating systems design. Incidentally, Microsoft Windows is also based on Unix. Thus, every piece of data, executable code, and even logged information was encapsulated in files. Files had the basic attributes -- create, append, write, delete, and read. From this perspective, any file including a system log file can be deleted.

## 6.4. Nature of Viruses

All of you are aware of application level problems. For instance, it is well known that Microsoft Windows and Outlook have significant weaknesses or security holes in them. Many of you, I am sure, have had the unfortunate experience of your computers being attacked by a virus [8]. Not too long ago, the laptop at our home in Arizona was infected by the NIMDA virus. My son helped me witness how the application files were slowly and automatically being deleted from the folders. I wish to draw your attention to the relatively slow rate of infection. The reason is that such viruses are executed relatively infrequently. We can assign a time constant to measure the rate, say it is in seconds, for the NIMDA virus. Now consider a virus that attacks not the high-level application programs but the low-level microcode that executes, frequently, in microseconds. This raises the possibility of an attack in the background that no one will even realize before the infection is total. Now, these things are not going to occur tomorrow, however, we must start thinking about them now, so we are not caught trying to play catch-up to the perpetrators.

Let me also discuss a little about viruses and intrusions. Many commercial companies have antivirus software. Let me present the facts and you can judge for yourself whether antivirus software is real. A virus is a string of 1s and 0s, it is in executable code form. When we logon to a website, we are allowing the distant server to download code in the form of a string of 1s and 0s onto our own machine. There is absolutely no way to distinguish between good code and bad, i.e., infected, code. So when we open a website, the virus comes straight into our machine, and we have cheerfully given pre-approval for this to occur. Beyond this point, there is nothing any one can really do to prevent the onset of the attack. When a virus comes and sits and does nothing (we tend to call it harmless) we cannot even detect it. When it starts to manifest itself, we then stand a chance to detect its presence if we are careful. Once we have identified its signature, we can take action should the virus attack again in the future in its original unmutated form. From nature, we know much too well how rarely a virus reappears in its original form. The fundamental difficulty with a virus then is that we must allow it to execute in order to detect it.

The real problem with viruses is far more severe. In fact, unlike today, viruses in the future will not even care to target personal computers; they will target routers and switches. The damage would be far more widespread and devastating. Let me present an example, a real story. If someone does not get to read a message that I had sent, one would think that is ok and little harm, if any, will result. Prior to World War I, the crown prince of the Austro-Hungarian Empire was assassinated. The Emperor sent a letter to the German leader, asking for advice and support for the war. The letter was delayed and, for some unknown reason, when the response arrived too late, the Emperor had already declared war on the Serbs. In the response, the German leader had advised against engaging in war. To the best of our knowledge, no one is known to have intercepted the message, read it, and suppressed it. It was simply delayed. Had the message arrived on time or had the Emperor not been so overconfident of German support and delayed his actions until receiving the official response, the story of the world would have been very different. I will submit to you that, sometimes, it is much more important for a message to get through, rather than being packaged with excessive security which may prevent it from going anywhere.

## 6.5. Third Key Element of Networking--Control Algorithm

The control algorithm mentioned earlier is what makes things work. We swipe our credit card at a gas pump; the information from my card is transported to a company that verifies that the card is not stolen and that there is enough credit remaining, and then instructs the pump to let me pump gas. This is an algorithm that is distributed and works in perfect unison. It resides partly in the gas station, partly elsewhere. But it is also a great source of vulnerability. Consider the recent terrorist bombing attack in Bali. When news first came out, it stated that there was a huge blast from a massive bomb that killed some 200 people. Later, the report was amended and we learned that there were two bombs. The first bomb, a smaller one, was blasted inside the club. It did not kill too many people but caused a lot of confusion. The panic stricken patrons ran out into the open -- an alley, looking for shelter, and then a much bigger, second bomb was exploded killing a lot of people. The terrorists knew they could not pack a large bomb in a small clubroom, so they conceived this sinister scheme. If we look at Nature that is exactly how humpback whales hunt for fishes. Once they spot a school of fish, a pack of humpback whales circle them, force them to tighten their formation, push them to the surface, and then devour them in large gulps.

Control algorithms are very powerful -- they can be great in their achievements or devastating in destruction. Consider another example of how control algorithms can be exploited. As many of you may know, the classic TCP/IP protocol helps ensure that a sender's packets reach the destination. Thus, if one or more packets do not get through, the sender retransmits them and continues doing so until successful. What would happen if a perpetrator examines a network and decides to intercept and delete a few packets? When the destination node sends a negative acknowledgment that packets X, Y, and Z are missing, the sender will retransmit, depending on its unit of storage, a group of packets that includes X, Y, and Z. In essence, repeated attacks will result in a large number of retransmissions, implying an overloaded network, possibly leading to failure.

Another example, a true anecdote from World War II, underscores the unprecedented power of control algorithms. After the occupation of France and prior to launching the air campaign against Britain, the German Air Force, the Luftwaffe, built two towers on the coast of France, directly across from Britain. There was very little activity in the towers and to most people, including British spies, there seemed to be no logic behind their existence. In reality, the towers housed two powerful EM signal generators of specific frequencies whose invisible beams were directed at Britain. Squadrons of German aircraft took off from different locations in occupied Europe headed towards Britain and the allies were astonished at the extreme precision of their bombing. At the time, inertial guidance system was the only known mechanism for navigation and it was relatively imprecise. Some of the German aircraft were shot down and the captured navigators could provide, if they volunteered, very little information. In reality, German navigators were provided two instruments, in essence EM frequency detectors, and given the following instructions. When the needle of the first instrument deflected off the scale, they were to open the bomb bay doors and stand ready. As soon as the second instrument went off the scale, they were to drop their payload. The two beams served to guide the bombers to their target with extreme precision, regardless of weather, or day or night. While even the German pilots did not know their targets until they dropped the bombs, their unpredictable flight paths deceived the British causing them to fail to scramble their fighters, in time, to intercept the bombers.

# 7. ENCRYPTION

Let us now turn our attention to encryption [4, 5] algorithms. While the Egyptians had hieroglyphics, the Chinese employed encryption algorithms, and the American Indians invented smoke signals. People have been using encryption for a long, long time. Then, in World War II, the Germans used it in a big way. If we carefully analyze the progress of World War II, we observe that the U-boat's use of the Enigma encryption algorithm posed a real challenge to the Allies. By rapidly sinking merchant and military ships, the Germans U-boats came perilously close to cutting off supplies of fuel, food, and ammunition to Britain and winning the war against the Allies. Military historians and analysts attribute winning the war against the U-boats to the Allies' successful breaking of the Enigma code. This is only partially true. First, the Allies never succeeded in completely breaking the Enigma code. Second, the real defeat of the U-boat came from a change of their command and control strategy, resulting from the German high command's desire to micromanage the U-boat fleet. This is described as follows. Every U-boat submarine was part of a wolf pack that was given very specific assignments prior to leaving port for a three-month mission. No one precisely knew their exact location at any given time. If it became necessary for the wolf pack to ask for instructions from the high command or if the high command needed to alter a mission, silence would be broken and the wolf pack would engage in a communication, still utilizing the Enigma code. Thus, communication was asynchronous and the Allies found it extremely difficult to locate them. When the German high command changed the command and control strategy and required the U-boats to transmit their positions regularly, the Allies were quick to note the precise time intervals, home in on their communications at those time instants using the directional range finders, triangulate the positions, and were literally on top of the U-boats minutes after they had started to report to the high command. The failure of the control algorithm was so catastrophic, that in a single month, some 70% of all U-boats were destroyed or sunk, independent of the Allies' ability to successfully break the Enigma code.

## 7.1. Formidable Challenges to Encryption

In the previous chapter it was noted that in the private-public key encryption mechanism, the most important element is the keys. It would appear that the best a perpetrator can do, is to guess the values of the keys. Thus, if one makes a key sufficiently long in size, the likelihood of a correct guess would diminish dramatically. There are two current trends that challenge this view. First, a perpetrator may attempt to break into the algorithm and obtain the keys by employing multiple computers to test different combinations simultaneously. Conceivably, one may utilize hundreds if not thousands of computers, processors, or IC chips for this purpose. In fact, a researcher at UC Berkeley had broken a code precisely in this manner, hours after the key was placed in circulation.

The second trend is the real killer. Most of these keys are based on prime numbers because it is computationally difficult to determine if a number is prime. For instance, to determine if the integer, N = 2561, is a prime number, one would have to resort to a brute force technique namely, determine if N is exactly divisible by 1, 5, 7, 11, etc. Recently, however, a research group from the Indian Institute of Technology, Kanpur, India, has invented a breakthrough algorithm that performs this task quickly. Thus, as algorithm designers also become adept at successfully utilizing faster parallel processing engines, encryption codes may be broken in a matter of hours, minutes, or seconds, instead of years.

## 8. ON THE RELATIVE IMPORTANCE OF DATA VS. INFORMATION

Clearly, encryption is applicable to both data stored locally in a computer or server and data while in transit. This has influenced our current thinking in a significant manner. Unlike in the past when access to privileged and secret information, stored at a site, was considered all important, today, greater importance is attached to information that is transferred back and forth between different users who continually add value as they process it. That is, the potential value of a piece of information, when permitted to be exchanged back and forth, can often be much higher than if it is simply locked up in a safe. To help explain this idea, consider a crude analogy. When a Mercedes or Lexus automobile dealer has lots of shiny new cars sitting in the lot, it looks good but the dealer isn't making much money. The dealer makes money when new cars come in from the factory and are sold out to customers. So, a successful dealer is one who has very few shiny cars sitting in the parking lot.

In the networking world today, on it's own, raw data has little value. Its value comes from what we do with it, how we process it. In this connection, let me present to you an anecdote, a factual story, about the value of data versus information. I was attending a conference on military command and control at the National Defense University in Washington, DC, several years ago. One of the keynote speakers was General Nelson, commander of the Ninth Air Force during the 1991 Gulf War and he made the following statement. He said, you know, if someone tells me that the enemy is 100 yards away from me, I really don't know what that means. Do I take out my rifle, my revolver, or do I run for cover? I don't have a precise feel for what to do with an enemy at 100 yards. But if someone tells me that the enemy is one football field away, I have a good deal of feel and understanding. Because I used to be a football player, I have an intrinsic understanding based on the speed I can run and other things, what a football field distance away really means. To me, this is information. If you say 100 yards, that to me is a piece of data for I cannot make use of it immediately. Sometimes, that can make all the difference. So, information is what we can make out of the data, the data per se, does not always have much value. Also, there is an element of personalization as information can mean different things to different people. Returning back to our original point of discussion, when data is being transported over the packet as packets, cells, or messages, its sanctity is of great concern.

## 9. SERIOUS CONCERNS WITH IP NETWORKS

Now let us discuss IP (Internet Protocol) networks. Most people have extensive experience with IP networks. I will share some radically new ideas and thoughts with you and apologize up front if I offend anyone. In my opinion, IP networks have very little future. The reasoning is as follows. If you wish to send email to someone, interactively communicate via a chat program, browse a website or download music or video, the IP is more than adequate. However, when you wish to send data or information, the compromise of which may have unforeseen consequences, or something that is extremely crucial and once lost, it is lost forever, such as medical or financial data, or trade secrets, that kind of data you cannot transport over IP and feel safe, ever. Someone can and possibly will break into it, today or tomorrow. The underlying reason is as follows. The IP protocol, an outgrowth of Darpa's effort - the Arpanet, is based on two fundamental principles. As long as these principles are in place, then there is nothing we can do to protect IP traffic, regardless of how many layers of software one sandwiches on top of it. The first is the classic store-and-forward principle, which I have already shared with you. Simply stated, this principle states, I don't care where the receiver is located, I will simply transport the packet to a node that I know and this node will then forward the packet to someone that it knows, and hopefully the packet will eventually reach its intended destination. In IP, there is no guarantee that the receiver will ever receive the packet. That's the number one concern. Number two is the end-to-end reasoning. This basically grew up hand-in-hand with encryption and states the following. I don't much care who, i.e. which nodes, lie between the receiver and me as long as there is some manner in which I can encrypt the packet that only the receiver, at the other end of the transmission, can decrypt the message successfully. We already saw this is not true. If a perpetrator sitting at an intervening node were to snap the communication channel, even if unable to decipher the message, the very purpose of the original transmission has being successfully attacked. In addition to these two key reasons, there are other serious issues with IP networks. For example, we had already discussed the problem of overload with the TCP-IP protocol.

## 10. THE FAA'S NETWORK OF THE FUTURE

I wish to share two interesting issues with you. Yesterday (2002), I was in Richardson, Texas, attending the ATM Forum's research meeting. The ATM Forum is the standards body for the asynchronous transfer mode (ATM) networking technology and is the counterpart of the IETF, the standards body for the IP networking protocol. First, we had a presentation from the chief architect from Harris Corporation whose team had just been selected to lead the multi-billion dollar effort over 10 years to completely upgrade the FAA network. He related the following. Some months back, the FAA had invited the NSA to run a diagnostic check on their network and assess its security, especially in the light of the 9-11 incident. During the course of routine interviews with NSA officials, a senior long-time veteran of the FAA made a remark. He said, you know, we have so much legacy software (legacy implying code that dates back 20 to 30 years) that no one really understands how it works. Without understanding, it would be difficult for anyone to break into it and so we are safe. Everyone at the meeting burst into laughter but you can visualize the dark lining beneath the humor. Now that is one way to ensure security, make sure no one understands what the code does, but that is not the logical way of doing things. Someone might and probably will figure out the legacy code and launch attacks. The worst would be that no one at the FAA will even have a clue about the attack. The second issue is that the architect mentioned they were committed to ATM technology.

# 11. NEW DIRECTIONS INTO SECURE NETWORK DESIGN IN THE FUTURE

So, what is the solution? Is there a new approach, is one even conceivable, or is the picture a dismal, hopeless one? Well, things are not all bad but we have to continue developing innovative solutions and never underestimate the enemy. Although a complete solution is not yet ready, we will present two key developments. The first is a framework that I had mentioned earlier and one that we had developed along with NSA. The framework allows us to do something unique and when we combine this with ATM networks [1, 4], what we get is a promising and realizable approach, at least one approach, to designing networks with the offer of reasonable security. What does the framework [1] look like? It appears as an innocent, two-dimensional orthogonal matrix, i.e. the rows and columns are independent of each other. If we examine closely, we notice that the term pillars is used for the entries along the rows. The reason for this is as follows. Security is analogous to a building where the pillars support the roof. In an ideal scenario, each and every one of the pillars is at 100% strength. In reality, however, given that security is prohibitively expensive and 100% security is unattainable in the absolute sense, while some pillars can and are rendered strong, others may be left weak. The combined strength of all of the pillars determines whether the roof will collapse or held in place. How does this simple idea hope to succeed in addressing the complex issue of network security? Very simply, the matrix constitutes an objective assessment, a practically realizable measure, of the quality of a network's security posture. This is its real contribution.

To gain a better understanding of the importance of the framework, consider the following example. Last year, we were scheduled to discuss an engineering design problem in the freshmen E 101 class. I asked the students to identify a problem from their everyday life. Mostly everyone complained that when walking on campus, his or her umbrellas often turn inside out in the strong wind. The problem is particularly aggravating when strong wind is accompanied by rain and they attributed the cause to Stevens' close proximity to the Hudson River. We engaged in a discussion and started to identify and address the issues systematically. We soon realized that the commercially available umbrellas lack any rating of the wind speed up to which they will function properly, i.e. without becoming damaged from turning inside out. We reasoned that if we conceived a new metric of an umbrella's effectiveness, manufacturers could use the metric as a standard, test their products under actual conditions, put labels, and price them accordingly. Thus, a consumer such as a Stevens' student may choose to pay an extra $15 for an umbrella rated at 30 miles/hr, confident that his or her expensive books won't get wet in the rain and that he or she would not have to spend an unknown sum of money during the winter months purchasing replacement umbrellas. In contrast, a person in inland Arizona would be quite content to buy a $5 umbrella since strong winds rarely accompany rain. Thus, the metric would not only help consumers make intelligent choices, but assist manufacturers in differentiating their products, possibly launching a new market of high performance umbrellas to discriminating customers. In much the same way, the security matrix will constitute a scientific standard in facilitating the security rating of networks, foster competition among industry to manufacture improved security products, and assist consumers including network managers in intelligent product choices.

The second key development is the integration of the security framework with an important attribute of ATM networks towards a promise -- a new approach to secure network design in the future. Recall the difference between IP and ATM networks. In IP, packets are simply forwarded in the general direction of the receiver. In contrast, in ATM, first a route must be established and, where successful, ATM cells are then propagated over the path. Thus, ATM is similar to classic telephony with of course some key differences. If we were to combine our security matrix idea with the route establishment process, also known as call setup, we can achieve the following. If a network fails to obtain a route such that every node and link along the route satisfies a user's security concerns, the call will fail, as it should, and the traffic will not be transported. Should the network succeed, the successful route obtained will impart a measured sense of trust and confidence in the user and the ATM cells may be propagated along this path. For more details, the reader is referred to [1] for details of this approach, termed security on demand, in the literature. Thus, the combination of the two elements succeeds in transforming a network's security posture into a measurable and usable quality of service metric, which future network designs may embrace as a practical approach to deliver security to its customers.

## REFERENCES

1. Sumit Ghosh, Principles of Secure Network Systems Design, Springer-Verlag (NY), April 2002, ISBN 0-387-95213-6.
2. H.J.(Jerry) Schumacher and Sumit Ghosh, “A Fundamental Framework for Network Security,” Journal of Networks and Computer Applications, Vol. 20, No. 3, July 1997, pp. 305-322. Academic Press.

3. H.J.(Jerry) Schumacher and Sumit Ghosh, "Unifying the Secure DoD Network & Public ATM Network Infrastructure," Proceedings of the MILCOM'99 Conference, Atlantic City Convention Center, NJ, Oct 31 - Nov 3, 1999.
4. Thomas D. Tarman and Edward L. Witzke, Implementing Security for ATM Networks, Artech House, Boston, MA, ISBN 1-58053-293-4. 2002.
5. W. Stallings, Cryptography and Internetwork Security -- Principles and Practice, Prentice Hall, NJ, 1999.
6. W. Stallings, High Speed Networking, Prentice Hall, NJ, 1997.
7. D. Bertsekas and R. Gallagher, Data Networks, Prentice-Hall NJ, 1992.
8. Sumit Ghosh, "Computer Virus Attacks on the Rise: Causes, Mitigation, and the Future," Financial IT Decisions 2002, Vol. 1, a Bi-Annual Technology Publication of the Wall Street Technology Association, Red Bank, New Jersey, http://www.wsta.org, Feb/Mar 2002, pp. 16-17, ISBN 1-85938-369-6.

**About the Author:**

A short biography occurs in the Preface. For further details, visit http://attila.stevens-tech.edu/~sghosh2 .

Chapter 3

# COMPUTER SECURITY: SOPHISTICATED ENEMIES

Yalkin Demirkaya
*Commanding Officer, Computer Crimes Investigation, IAB, NYPD*

**Abstract:** This chapter opens with a brief overview of the security threats faced by industry and our lack of preparedness against sophisticated cyber adversaries. It then classifies the types of individuals that threaten our security and describes the tools they use to perpetrate crimes and perform acts of industrial espionage. It ends with an overview of the techniques we can use to guard our organizations against cyber attacks.

**Key words:** Computer Security, Sophisticated Enemies

## 1. THE PROBLEM

"Most companies don't spend as much money on protecting their data as they do on coffee for employees. Less than 0.0025 percent of corporate revenue is spent on corporate information-technology protection. ...Our adversaries, be they run-of-the-mill hackers or devoted members of terrorist cells, have the same training and much the same access to technology as we do. Our future enemies understand our technology at least as well as we do... Most of the nation's critical infrastructure--the power grid, voice networks, and water supplies--are vulnerable. You'll find computers at the heart of all these systems. Terrorists have a wide range of technology targets, not all of them in cyberspace." (From a speech by Richard Clarke, White House Special Advisor, on Cyber Security Issues [2].)

The above statement very simply and eloquently outlines the problem. When we add total lack of awareness and understanding by most corporate executives, and the presence of sophisticated adversaries, the picture becomes very grim. Consider the following statements and replace the "X" with your company name to see how accurately this description represents your organization:

- X lacks the analytical framework necessary to address insider threats.
- Inadequate physical protections place electronically stored information at risk of compromise.
- X lacks adequate, documented INFOSEC policies.
- Confidential information stored on widely utilized systems is not adequately protected.
- Many key Information Security positions remain vacant and, when they have been filled, the people assigned often lacked the time and authority to perform their duties.
- Some X systems have insufficient resources to perform required audits. When audits are performed, the audit logs are reviewed sporadically, if at all.

The above statements are part of the testimony given to U.S. Congress by Attorney General John Ashcroft on the current state of computer security in the Federal Bureau of Investigation [1]. If the premier law enforcement agency responsible for fighting and investigating computer crime is in this shape, one can imagine how well the rest of our infrastructure is protected. We can probably replace the "X" in the above statements with the names of 95% of the large corporations and most government agencies in the United States. Former Special Agent Hanssen spied on the FBI with impunity for over twenty years using various data hiding techniques that no one in the FBI was capable of detecting. There were no checks and audits against internal threats in the FBI at the same time that the bureau was telling the entire country year after year through their surveys with the Computer Security Institute (CSI) that the real computer security threats originate from within the organization.

This chapter will deal with computer security issues in the post 9-11 world—a world made even less secure because there are countries and groups of people in the world who want to destroy us. Recently the law enforcement community got their first look at the al Qaeda training manual. This manual apparently borrowed many concepts from a publication called "Unconventional Warfare," which still can be found on the Internet. According to both documents, the only way to defeat the United States is to attack its economy. So our enemies are strategizing on how to defeat us by economic means. One of the most effective ways of doing this is industrial espionage. When we add to the equation that most of our military allies are our industrial rivals, we can see the urgency of the situation.

Unfortunately, very few of our organizations are currently prepared to deal with the level of threat that is posed by sophisticated adversaries. I do not know any private corporation that has a dedicated group of people whose sole responsibility is to detect and investigate insider misconduct both reactively and proactively. Any company can, and most will, respond to an incident after its occurrence is discovered, but the real challenge lies with discovery and detection before an incident occurs. It is my personal opinion that the vast majority of security breaches go completely unnoticed when committed by sophisticated insiders.

The annual report to the Congress on foreign economic collection and industrial espionage compiled by the Office of the National Counterintelligence Executive estimated the total economic loss for the year 2002 to be as high as $300 Billion and these losses are rising each year [4].

To illustrate the level of threat that this country is facing let us look at a single criminal case investigated by the FBI (http://www.fas.org/irp/ops/ci/docs/fy98.htm). As a result of a single industrial espionage activity, the trade secrets of a US company were stolen and transferred to a foreign company. This resulted in the following:

- The foreign competitor captured the international market
- The U.S. business lost $600 million in sales
- An estimated 2,600 jobs were lost
- An additional 9,542 jobs were estimated to be lost during the next 14 years
- The U.S. trade balance was negatively impacted by $714 million
- The lost tax revenue was estimated to be $129 million

No country spends more money on research and development of new technologies than the United States. Our businesses are competitive in the world market where they have the edge not because they can produce things cheaper but because they can produce things that no one else can produce better. As reported by the National Counterintelligence Center (Annual Report to Congress on Foreign Economic Collection and Industrial Espionage - 2002) the following are some of the industries targeted by other nations and organizations: information technologies, lasers, telecommunications, manufacturing, biomedical technologies, composite materials, alloys, advanced electronic devices, aerospace structures and propulsion systems, directed energy and kinetic energy applications, etc. The list is ever changing and can fill many pages.

In addition, we depend on technology in our everyday life more than any other country in the world. Computers run every aspect of our life from communications to finance to energy and transportation. A successful attack on our computer networks can deliver a devastating blow to our economy. Yet, an average corporation spends more money on food and beverage for its employees than they spend on IT security. The law enforcement community got a wake up call on 9-11. We no longer feel invincible within our own borders. Until the IT community gets its wake up call, I believe the current trend of not allocating enough resources to security will continue.

Currently, almost all IT security efforts are concentrated on stopping low level threats. For most organizations, virus protection or intrusion prevention by "script kiddies" are the primary concerns. Very rarely can one find an organization that has proper defenses against internal or external threats that are posed by sophisticated enemies.

This chapter will only concentrate on the sophisticated adversaries, both internal and external. We will talk about who they are and what they are capable of doing. Sophisticated operatives pose two main high level threats: espionage and sabotage. When dealing with hostile organizations such as al Qaeda, and states that provide support for such organizations, we have to worry about both classes of threat. A number of war games conducted on cyber terrorism issues have revealed that, in each scenario, the enemy needed an insider to inflict the most damage. To illustrate the severity of the problem, consider the chemical plants in the United States that produce pesticide. A single incident at a Union Carbide plant in Bhopal, India perpetrated by a disgruntled insider caused 3,800 deaths and over 11,000 injuries. The questions I pose are: How many similar plants are operating in the United States? How computerized are they? And how many of our organizations have excellent computer and physical security?

## 2. MEET THE ENEMY

The following are seven different classes of individuals that are a threat to an organization's digital assets:

**Script Kiddies:** These are individuals with limited knowledge who obtain readily available tools from the Internet to cause damage to other's property. They are no different than a teenager who buys a can of spray paint from a hardware store and defaces someone's property with it. The teenager does not possess the knowledge to manufacture the spray paint but will use it to cause mischief. Script kiddies are a low level threat but can be very annoying or damaging if the target is an organization that conducts business on the Internet. Their motivation could be anything from political activism (hacktivism) to peer recognition. They are relatively easy to stop and apprehend.

**Black Hat Hackers:** The term "black hat" is used to describe a hacker (or, if you prefer, cracker) who breaks into a computer system or network with malicious intent. Unlike a white hat hacker, the black hat hacker takes advantage of the break-in, perhaps destroying files or stealing data for some future purpose [5]. The 13 year-old hackers reported in the media do not exist. The 13 year-old has not lived long enough to acquire the level of knowledge and/or skill to become a real hacker. Hackers are generally much older with extensive knowledge in various operating systems and programming languages. They are largely motivated by their love of computers and live for peer recognition. Generally, they are extremely skilled and intelligent. These are the people who discover all the weaknesses in different operating systems and software. They write computer programs to exploit these weaknesses. They then publish their accomplishments on the Internet to gain peer recognition. Unfortunately, script kiddies often download the software programs produced by hackers and use them on various systems to cause mischief.

**Disgruntled Employees:** The damage caused by disgruntled employees can be tremendous. Publicized cases of security breaches by disgruntled employees usually involve millions of dollars in damage to the organization. Most often the employees involved are IT personnel. Disgruntled employees can cause extreme damage if not handled properly. They are relatively easy to catch if the situation is handled properly to prevent damage and preserve evidence for possible criminal or civil action.

**Cyber Criminals:** Members of this group are criminals in a traditional sense who usually employ the talents of hackers or the services of script kiddies to extort money from large organizations. More often than not they will penetrate an organization's IT defenses and steal information. The stolen information is then used to extort money from the victim.

**Computer Experts:** Unfortunately, certain people in the computer field feel that they know everything about all aspects of computers and give this impression to their employers. Because employers lack the knowledge to question the level of expertise of their employees, they often take bad advice and make grave mistakes. The computer field is far too complex to be mastered by any one individual. This is why we have programmers, network administrators, communication specialists, etc. Taking computer security advice from a self-proclaimed IT expert is like going to an internist for heart surgery. Frequently, organizations make bad decisions and fall victims to computer crime because of the inappropriate advice they receive. IT experts do not exist! Sometimes there is nothing more dangerous to an organization's digital assets than a self-proclaimed computer expert who knows it all.

**Professional Spies:** Many countries that are our military allies are also our industrial enemies. Why spend billions in research and development when you can get better results by obtaining someone else's research? This philosophy makes very good sense to some very large corporations. For example, when Mitsubishi corporation decided to enter the space industry in 1986, they took advantage of our Freedom of Information Act laws and filed over 1500 FOIL requests with NASA in 1987 alone. In an era when almost all of our trade secrets are stored in computers, it becomes relatively easy to employ an insider to steal that information. If inappropriate security measures are implemented, it is relatively easy to obtain the most closely guarded information without detection. Professional spies do not get much press but are the biggest threat to any industrial or government organization. They can be stopped with proper measures and diligence. If their existence is suspected, with the deployment of counter measures, they can be detected and their employers identified.

**Enemy Operatives (Terrorists):** Being able to cause damage to your enemy without any physical danger to yourself is very appealing to the enemies of this country. Our enemies are very motivated and well trained. Their mission is to cause as much damage as possible by any available means. Their primary objectives are espionage and sabotage. The al Qaeda manual mentioned above directs their members to be patient. It advocates surveillance from within. Terrorists are actively trying to exploit cyber space to cause harm to us.

In the remainder of this paper, our discussion will address the problem of enemy operatives and professional spies. Both groups may have the resources of foreign governments or very large organizations at their disposal. Very few organizations are prepared to deal with them. I have yet to see an IT security system in any company, which can detect and capture a professional spy who is an insider. Almost always professional spies and enemy operatives are discovered by accident. One of the fundamental challenges of computer security is the discovery of wrong doing in the first place. If someone comes to your house and removes a valuable item, you will know immediately because the item will no longer be there. On the other hand, if an insider copies your most valuable trade secrets from your computer system, you will never know because the data will still be there when you look for it. Because almost all of our IT defenses are designed to stop unauthorized outsiders, we rarely have any protection against an authorized insider engaged in an unauthorized activity.

## 3. TOOLS OF THE TRADE

What are some of the tools used by the individuals and groups listed in the previous section? In this section of the paper, I will give a simple overview of some of the tools and techniques used by our adversaries. I believe awareness is the most important factor in countering these threats. The purpose of this chapter is not to make you a security expert, but rather to make you aware of some basic techniques for penetrating the security of your organization. With this information you will be better prepared to recognize a potential problem when you see one and, hopefully, to get help from qualified individuals who can properly manage the threat.

**Social Engineering (Sophisticated Operations):** Most successful hackers possess more social engineering skills than technical knowledge. They are great conmen who have some technical knowledge and a God-given talent to convince people to do things they want done. Social engineering is the most effective way of breaking into computer systems.

To illustrate this threat, let us create a sophisticated operation to target a single individual who is going to get us into where we need to go. As a detective commander who investigates computer crime, I have conducted many sophisticated sting operations that are best described as very elaborate social engineering schemes. The objective is to have the target do something that you want done. If I target a secretary in your organization, with a team of 6 skilled individuals for over six months, I am sure I can have him or her receive an e-mail from me with a hidden payload. That payload will create an outbound connection from your organization that, most likely, no one will notice. (Most organizations do not log outbound connections on their firewalls and many never even examine their firewall logs at all.) This connection will give the individual complete control over a machine in your network from where he or she can run other software to gain control over other machines.

I had one operation where the target of our investigation communicated with four of my detectives for over five months without guessing who they were. There were times when this individual entered a chat room in which he was surrounded only by my detectives. No matter who he was chatting with, he was talking to one of my investigators. Obviously, we studied him so well that we only told him what he wanted to hear. During the first few months we just gained his trust. In the fifth month he was committing an average of five serious felonies per day online. Incidentally, the target of our investigation was a police sergeant with 14 years of law enforcement experience. If an experienced law enforcement official can be fooled this way, think what can be done to an average employee in your organization. The point is that, sophisticated social engineering attacks are extremely difficult to prevent unless every member of the organization is trained to deal with them.

**Data Hiding:** In general, data hiding is done in two different ways. One way is to utilize steganography software and the second method is to use various DOS tricks. Steganography software will allow a computer file to be hidden inside another file. For example, a spreadsheet that contains sensitive information can be hidden inside an image and can be e-mailed out of the organization's network. Spies do not have to make secret drops in remote locations anymore. All they have to do is put the information inside an image file such as a picture of a 40" plasma display and post it on ebay for sale. It would be very difficult to determine who the intended recipient of the hidden payload is when you consider the item may be viewed by a couple of thousand people. Most of the time no one will even know about the hidden payload to begin with and tracing the recipient is an even more difficult challenge. DOS tricks, such as creative use of the "copy" command with proper switches, are in the arsenal of most sophisticated operatives. By utilizing the "copy" command, a piece of text file can be added to the end of a graphics file, which would make it very difficult to detect. Also, by using disk editors such as Winhex and Norton Disk Editor, a floppy may be rendered unreadable effectively hiding its contents and making it appear to be an unformatted disk.

**Password Breakers:** Unless used intelligently, encryption can be detrimental to an organization's security if it promotes a false sense of security. There are many different password recovery software suites that will break all but the most sophisticated passwords. Password hashes are attacked with a disturbing consistency by using password recovery tools. Most application passwords are breakable very easily. An average Pentium 4 based PC can break a 6-digit password consisting of just numbers in under 4 seconds by a brute force attack. Windows server administrative passwords can be attacked with replacement methods where the secret password hash is replaced with a new password chosen by the attacker. Software such as LC4 and Jack the Ripper perform dictionary and brute force attacks on Windows and Unix network passwords. It is my experience that LC4 software will compromise about ten percent of the network passwords in most organizations within five minutes and another twenty-five percent within one hour. If the software is allowed to run for a day, more than eighty percent of all user account passwords will be compromised.

**Spyware:** There are two types of spyware: software-based and hardware-based. Within each category, there are different levels of sophistication. Hardware spyware that is readily available on the market at relatively low cost involves keystroke capture devices that go between the CPU and the keyboard. There are also versions that are keyboards themselves. These devices are capable of capturing up to two million keystrokes and will store everything that is being entered into a particular computer. They can only be detected by physical inspection. These devices are available for purchase on the Internet at a relatively low cost - most for under a few hundred dollars. Software-based spyware can be just as deadly if deployed properly by an insider. For under one hundred dollars one can purchase a very sophisticated spyware system that will stealthily capture every screen, keystroke, etc. These programs are so sophisticated that they are very difficult to detect and there is no silver bullet approach to detecting them. The only sure way of checking to see if spyware is deployed is by a manual examination of the suspected PC by a qualified technician. There are some anti-keylogger software programs on the market but our testing found that they are not very reliable. Up to this point, we are talking about commercial grade spyware. Any one who is familiar with programs like "Spector" know that a hundred dollars can buy you a quite sophisticated piece of spyware. Well how about spending twenty-five thousand dollars? There are government-grade spyware systems that are both hardware- and software-based and that sell for close to six-digit figures. For obvious reasons, I will not go into the details of what they can do, but you should be aware of their existence. Unlike encryption technology, there are no export restrictions on spyware. According to manufacturer policies, spyware software that is produced in this country is only sold to law enforcement agencies either foreign or domestic. Noting that some of our military allies are also industrial enemies, there is a possibility that U.S. produced spyware purchased by the national police of another country may be deployed against some of our own domestic companies. Industrial espionage is big business and when we consider that all of our propriety information is stored in computers, availability of these hardware devices and software programs should make anyone who is charged with safeguarding an organization's information assets very nervous.

**Computer Forensic Tools:** Generally when we talk about computer forensics, we are referring to evidence recovery from certain areas of disk drives. There are areas on disks such as file slack, ram slack, windows swap file and unallocated space where data is stored by the operating system that are beyond the reach of the user. Some computer forensics tools are used to retrieve data from these areas. The amount of information that is stored is enormous and can be of any imaginable type. Network forensics involves data flowing over the network that can be captured recorded and analyzed by network forensic analysis tools (NFAT). These tools operate as sophisticated sniffers and can be deadly if deployed by an insider against the organization. Forensic tools can be used either offensively or defensively. On the one hand, the capabilities of these tools are so enormous that they can be a tremendous threat to any organization's digital assets. On the other hand, computer forensic tools, if used properly, can be a great asset in detecting and investigating insider misconduct.

**Traditional Means:** There are known instances where foreign companies have offered huge bounties for laptops belonging to senior members of their competitors. On one occasion, a quarter of a million dollars was offered for a laptop who's owner was away in the Middle East bidding on a billion dollar contract. It is rare that the data on a laptop is protected by strong encryption to make the contents useless in case of a theft. We hear about laptops being stolen every day at the airport, hotels, etc. One cannot help but wonder how many of those laptops were stolen because of the data they contained. There are forensics hard drive duplicators that can be attached to a laptop in a hotel room with the ability to transfer the entire contents of a hard drive in under five minutes. It is not very difficult to gain entrance into someone's hotel room in a foreign country to take a copy of the hard drive. How many times are employees challenged when they hook up a digital camera to a company workstation to show family pictures to co-workers? Usually, there is no suspicion of any wrongdoing. Today a digital camera with a 2 GB compact flash card can take a copy of the entire contents of a research and development server in a highly classified environment. If the workstation is running Windows XP, any kind of a USB device including MP3 players, which often have storage capacity of over 40 GB, can be attached without the need for any special software or drivers. The device can act as an external hard drive on which the contents of the computer's conventional hard drives can be copied.

Foreign governments and large foreign corporations have people specially trained to steal our trade secrets so they can defeat us on the battlefield of industrial warfare. The spy schools that trained countless KGB agents and operatives of former Soviet Block countries are still in business today. The only difference is that the schools are open to anyone who is willing to pay the tuition. Ideology used to be the driving force behind schools like this. After the fall of the Soviet Union, cash replaced ideology as the driving force. These schools teach courses such as agent recruitment and handling, signals intelligence, etc. Their graduates are probably seeking employment with the strategically selected US companies who compete in the global market. Noting that Ford Motor Company reportedly spent 3 billion dollars in research and development of the Ford Taurus passenger sedan, the importance of industrial espionage becomes very apparent.

There are numerous occasions in which the agents of foreign governments were discovered entirely by accident. For example, there was an incident in which a French Intelligence service recruited several French nationals who were working at the France-based offices of Texas Instruments. These agents were told to steal information on marketing plans, product specifications, trade secrets, etc. French Intelligence passed the collected information to competing French companies including Machine Bull. In 1993, Bull sued Texas Instruments over patent infringement on a computer chip. Texas-Instruments later discovered that Bull originally stole the design from them through an agent who worked for Texas-Instruments for 13 years. The Intelligence Threat Handbook that is published by Interagency OPSEC Support Staff [3] is an excellent reference to give the reader some insight about the level of threat our companies are facing.

## 4. WHAT WE CAN DO

Most IT security threats can be defended against successfully. However, it is imperative that the proper resources are allocated and that a commitment to IT security is made by corporate leadership. Let us take a look at possible defenses that can be easily deployed against internal and external threats. Many books are written on the subject of computer security. Here, in a very limited space, we can only provide a very simple and crude approach to stopping some of the high level threats.

We will discuss external threats first since they are the easiest to stop. The first approach is to isolate critical information from the outside world. A computer system that houses an organization's trade secrets should never be open to the Internet. Proper use of "air gaps" will solve the problem of external intrusions very simply and effectively. I run such a system and do not even allow floppy or compact disk drives to be present on workstations in the network. The network consequently operates without any outside connection. I would be tempted to shoot any one who even comes close to this network with a telephone line. In this situation all I have to worry about is insider threats that we will deal with separately below.

Systems with less critical information that must be open to the Internet should be secured by regular testing of the borders against known vulnerabilities. Generally, it is advisable to employ an outside firm that specializes in such services. All recent high-profile cyber security incidents have taken advantage of vulnerabilities that were well known and where fixes were available for months. Organizations that fell victim to Internet worms such as Nimda, Code Red, Slammer, etc., either did not know they were vulnerable or never bothered to apply the patches that were available. As far as I am concerned, IT staff of these organizations were either incompetent or the management ignored their warnings and recommendations. Constant testing of borders and fixing the discovered vulnerabilities will solve 90% of the external intrusion threats in an environment where firewalls are configured properly and effective intrusion detection systems are deployed. Proper management of e-mail borne threats such as trojans, and proper training of the employees including the executive corps and the IT department should solve 98% of all external threat problems. Remember, 100% computer security cannot be achieved. Also, that an intrusion detection system that generates dozens of false alarms everyday will be ignored and will be useless after a few months.

Internal threats to an organization's digital assets are far greater and far more difficult to detect and stop. With very few exceptions, no organization is prepared to deal with internal threats and almost none employ proactive measures to detect unusual activity. Computer forensics tools can be used very effectively against insider threats. Network Forensic Analysis Tools can be very effective in identifying and documenting suspicious activity. At least one of these tools can use n-gram analysis (a technique to determine relationships between similar types of information) to quickly identify suspicious content and gather required evidence very quickly and effectively. Random and targeted spot checks of computers using forensic tools can be a great proactive measure to deter insider misconduct.

The research and development section of the company must have a separate computer network with diskless workstations that are completely locked down with no outside access. Proper use of encryption can be a great asset in safeguarding the trade secrets of the company. But, one must be mindful that the best encryption tool in the world can be defeated relatively easily if the environment allows the encryption keys to be stolen using a twenty dollar spyware program. Company computer usage/security policy should never allow personally owned laptops to be connected to the company network. Special attention must be paid to the connection of external devices such as digital cameras, MP3 players, USB bus storage devices, etc., into workstations on the company network. Routine checks against spyware/trojans should be conducted on sensitive systems. And, most importantly, the members of the IT department should receive proper training on how to detect and react to insider misconduct. If warranted, dedicated people with proper training should be employed to deal with insider IT security issues. Remember, the insider you are going after may be a member or even the head of the IT department. You will never know who you may have to go after. There is a well-known case where a CEO of a major U.S. auto manufacturer stole the trade secrets of his organization and took them with him to his new employer, a well known German auto manufacturer.

# 5. CONCLUSION

In conclusion, it is unfortunate that our IT defenses are no match for the sophisticated adversaries that are targeting our companies, our economy and our country. The level of naivity on the part of the executives of large corporations is mind-boggling. CEO's and CFO's are often unwittingly the best allies to the enemies of this country and their own organizations when it comes to IT security. They are the ones who operate on a "what am I getting for my investment" principle and spend more money on coffee than securing their IT infrastructure and digital assets.

I have yet to meet a CEO who is not convinced that his IT people are competent to safeguard the information assets of the organization. Remember, all of these companies who fall victim to well-publicized worms had the best IT department in the world staffed by bona-fide computer geniuses. Unfortunately, with the exception of a very select few, most IT departments do not have the necessary knowledge or the resources to deal with even a low-level cyber threat let alone a sophisticated insider. The only way to deal with the level of threat posed by sophisticated adversaries is proper training and awareness. Most IT security threats can be defended against with relative success. However, it is imperative that the proper resources are allocated and commitment to IT security is made by the people who are running our corporations. Investment in IT security and protection of trade secrets unfortunately can't be measured in dollars earned. Nevertheless, every organization should engage in a serious risk assessment exercise and allocate resources sufficient to mitigate exposure to large losses.

# REFERENCES


1. Ashcroft, John (2001). Reported in U.S. Department of Justice, A Review of FBI Security Programs, Commission for Review of FBI Security Programs, March 2002

2. Clarke, Richard (2002). Reported in David Coursey, "Cyberterrorism - the new Cold War," ZDNet UK, 22nd February 2002

3. Interagency OPSEC Support Staff. http://www.ioss.gov/

4. Office of the National CounterIntelligence Executive (2000) http://www.ncix.gov/

5. SC InfoSec OpinionWire, Wednesday, December 11, 2002 http://www.infosecnews.com/opinion/2002/12/11_01.htm


**About the Author:**

Mr. Yalkin Demirkaya has seventeen years of law enforcement experience as a Detective as well as a Detective Squad Commander. He possesses twenty-two years of computer experience as a white hat hacker. He is the founder and currently the Commanding Officer of the Computer Crimes Investigation Unit of the New York City Police Force—one of the largest law enforcement organizations in the world. He is also the president of Digital Services Co., which specializes in providing a vast array of computer security related services to the business community ranging from vulnerability assessments to detection of insider threats.

Mr. Demirkaya has been responsible for the formation of policies and investigative procedures in the area of computer crime. He has lectured extensively and has provided training in the area of computer crime to the law enforcement community. Mr. Demirkaya holds a B.S. in Police Science and an M.A. in Criminal Justice and is a Certified Computer Forensics Examiner.

Chapter 4

# DIGITAL PEARL HARBOR: A CASE STUDY IN INDUSTRY VULNERABILITY TO CYBER ATTACK

Eric Purchase and French Caldwell
*Gartner Inc.*

**Abstract:** This chapter describes a war game conducted by Gartner Inc. in July 2002. Security experts and the US Naval War College in four different industries were asked to plan cyber attacks on their own industries. The results of this gaming exercise are explained and conclusions are drawn concerning the extent of our exposure to concerned attacks by cyber terrorists.

**Key words:** Digital Pearl Harbor, Cyber terror, Strategic Planning

## 1. THE ROLE OF COMPANIES IN DEFENDING AGAINST CYBERTERROR

Participants in a war game called Digital Pearl Harbor sponsored by Gartner and the U.S. Naval War College demonstrated that terrorists could use cyberattacks to hurt the U.S. economy and political will severely. The "Digital Pearl Harbor," war game occurred on 24-26 July 2002 at the U.S. Naval War College in Newport, Rhode Island. The 80 some players were drawn primarily from Gartner's client base and included people who manage the IT systems that are part of the critical infrastructure, that is, systems that control or operate parts of the electrical power grid, telecommunications, financial services, and networking services and the Internet. A multi-industry strategic war game centering on cyberterrorism had never been played before.

The war game took the form of a structured workshop to explore how an enemy — the "Red Team" — might plan an attack on the United States. The creators of the war game set participants the goal of using cyberattacks to cripple U.S. economic and national infrastructure to the extent that it shifts the balance of power. As the focus for Digital Pearl Harbor, the creators chose four industries where a successful attack on any one of them by itself would cause substantial damage, but the four together seemed likely to produce a coordinated set of attacks that would inflict maximum damage. The industries were telecommunications, the Internet and networking services, electrical power and financial services.

Participants in each industry group, or "cell", worked together in rooms that had a computer on a LAN connecting the four cells and a fifth cell, the Red Team command cell, which coordinated the efforts of all four groups. The participants were limited by very few baseline assumptions. However, the game rules provided some basic boundaries

1) The attackers were terrorists rather than hostile governments.

2) They worked with the technologies and infrastructure deployed today.

3) Each of the four groups had fixed resources to work with, such as no more than $50 million, five core members and limited physical attacks to complement the cyberattacks.

4) They could have no more than eight physical attacks, and those physical attacks had to create a synergy so that the effects of the physical and cyber attacks together would be much greater than either by itself.

A Gartner moderator guided the work of each of the four groups, and a few "White Team" members monitored the discussions to ensure that the "Red Team" attackers made realistic proposals.

The Red Team commander sought to minimize conflicts between the four cells. Two or three representatives from each group would meet periodically with the Red Commander, who oversaw the creation of an overall timeline for the attacks. For example, the Internet group would need the telecom infrastructure for its own attacks, so the Red Commander ensured that attacking the telecom system came later in the overall attack scenario.

The war game unfolded in four segments, which each group tackled individually:

1. Strategic planning — what goal the group should set

2. Preparation and reconnaissance — determining what targets could best be attacked to achieve the group's goal and what resources would be needed

3. Attack and counter-recovery — a plan detailed enough to show the feasibility of the proposed attack

4. Assumptions and changed premises — examination of the assumptions behind the attack plans and revision of them to achieve the goal within two years (or six month in the case of the Internet group.)

Following each game segment, all groups met together to present their results and to question and discuss. Let's look a the findings of the individual groups before drawing some general conclusions.

## 2. UTILITIES

The war game revealed that no single cyber attack on the control systems of the nation's electrical power grid will likely cause anything other than an isolated outage. Terrorists can't just sit in front of PCs somewhere and bring down the U.S. power transmission system. For more than a small, short-term power outage, the electrical power cell proposed "social engineering" and physical attacks to abet the cyber attack.

As yet, utilities have deployed little security technology on control systems infrastructure to thwart attackers. However, the control systems infrastructure—supervisory control and data acquisition (SCADA), substation automation and so on—increasingly uses industry-standard IT solutions. Therefore, utilities must think about securing their systems the same way that traditional IT systems are secured.

### How the war game played out

The attack planned by the power group closely resembled what cyberterrorism against the electrical power grid would probably look like because it focused on a few markets, with the intent to create secondary cascading outages throughout the larger region of the grid. There are four major grids for North America: Eastern, Central, Western, and Texas. The cell took a two-pronged approach: a physical attack against up to 24 transmission towers at key points in three markets as well as a cyberattack against three to six SCADA systems in the same markets.

The planners aimed first to cause confusion and uncertainty, then to disrupt power in key markets and, finally, to disable key control systems for extended periods. These attacks would amplify the effects of the attacks planned by the other industry groups (telecommunications, the Internet and financial services) by eliminating the power for other parts of the infrastructure. For example, even if a telecom could regain control of the hacked routers, it still would have a hard time getting more than back-up electrical power thus leaving many of the routers out of service. The power group's ability to support the other groups led it to alter its initial plans, which called for attacks in the summer, when power grids face the greatest strain. Instead, the power grid attacks were designed to occur near the beginning of winter and the holiday season.

## Remote Attacks Are a Remote Possibility

The Digital Pearl Harbor war game showed a particular idea to be a myth: the belief widely held inside and outside the power industry that critical systems, such as SCADA devices and subsystem automation, can be easily accessed remotely so as to hurt the reliability of the entire electric grid. In reality, most of these systems remain isolated from the Internet although this is rapidly changing as more standards-based systems replace older technology, and secondarily, as SCADA systems are linked to corporate LANs for business reporting purposes. For the cyber attack, the cell determined that the best means would be social engineering to gain access to SCADA systems. For instance, the cell postulated that they could acquire a company that does maintenance on SCADA systems as a means to gain access and insert trojans into the SCADA software. The Internet cell however said that it would find SCADA's that are connected to the Internet during the reconnaisance phase of the operation -- so both social engineering and Internet hacking could be used in the SCADA attacks.

## IT Runs Your Operations

Power companies have done a good job of preparing for natural catastrophes, such as ice storms, and of adjusting to power shortages.

However, the industry must improve its vigilance against terrorists and other cyberattackers who can do great damage with focused, determined attacks. Power companies have a real and growing exposure that many in the industry have been slow to understand. They have deployed only the most rudimentary security services and devices to protect the command and control elements. Utility engineers oversee most operations, and they instinctively discount the threat from software and code, as long as the electromechanical systems and proprietary systems are safe.

Enterprises today count on obscurity to keep SCADA systems secure. Five trends show this confidence is ill-founded:

• Control networks continue to move from analog and electromechanical technology to digital technology. Utilities use the public telecommunications network to reduce the cost of operations.

• Public telecommunications equipment tends to converge on industry standards and to move away from proprietary specifications, which means many more people now understand the control systems power companies use; therefore, the pool of potential attackers has increased.

• The need in deregulated markets for real-time information about the state of the grid makes has led some power companies to close the "air gap" between the corporate IT systems and operational control systems . Furthermore deregulation has created more players in the operations of the electrical grid, all of whom need what used to be internal corporate information to do their jobs. Thus, the need has grown to publish, widely and freely, more information about the state and condition of the grid.

• Security measures within the SCADA and command and control elements of the nation's utilities vary widely. Most utilities don't put SCADA and operational control systems security under the same hat as IT security. Much coordination of security and sharing of security information occur by chance.

• Deregulation of the power industry has caused utilities to be more cost-sensitive. In a regulated environment, they could easily pass on the cost of enhanced security to consumers. In a deregulated environment, no one wants to be the first to create a more-expensive infrastructure. For this reason, guidelines and policies need to be enunciated so that all utilities act consistently.

## Recommendations

• Power companies depend on SCADA to control their operations and should, therefore, provide security in proportion to the threat. Security measures for SCADA systems should include intrusion detection and strong authentication measures, as well as recovery plans for the operation centers and SCADA systems.

• Power companies should add protection to transmission corridors through remote areas used to move large blocks of power between regions.

• The federal government must also confront this transmission issue to improve homeland security. As a first step, it should provide the industry with a set of consistent and minimum standards, followed by guidelines on access to information, physical access, and security measures and processes for the command and control elements of the nation's electrical infrastructure. An appropriate mechanism to make these standards and guidelines enforceable, consistent and realistic is voluntary government/industry interaction. For example, the National Security Telecommunications Advisory Committee (see www.ncs.gov/nstac/nstac.htm) offers a model for what the utility world needs.

• The federal government and the utility industry must clearly understand the importance of the command and control elements of the utility industry and not expose these elements to unnecessary risk solely in the interest of cost savings. The industry must adopt a set of security objectives and standards that continue to mature as threats evolve.

# 3. FINANCIAL SERVICES

Because the financial services industry encompasses many different types of businesses and uses widely divergent technologies, the war game participants came from various industry segments and were not accustomed to thinking in strategic terms that represent the total industry. The participants were technicians and managers overseeing individual systems or business processes. They hadn't deeply considered how cyberworld systems contribute to the overall working of the U.S. financial system. Yet, the variety of the technologies used throughout the financial industry makes it more difficult to do widespread damage. Terrorists would be challenged to understand the total infrastructure supporting one financial service segment, let alone the variety of segments that create the converged industry.

## How the Game Played Out

The financial services cell broke down the industry into three parts — cash flow, credit and infrastructure — and three markets — consumer (people), corporate (industry) and institutional (financial markets). The group then developed a list of targets for each part of each market (see Figure 1).

| | Cash Flow | Credit | Infrastructure |
|---|---|---|---|
| *People* | *Bank Accounts* | *Credits Cards* | *Investment Future* |
| Industry | Sales and Receivables | Credit Facilities | Infrastructure |
| Financial Markets | Settlements and Transactions | Credit Facilities | Depositories and Other Markets |

Figure 1. Matrix of attacks planned by financial group (source: Gartner Research).

Next, the group divided into two. One subgroup planned attacks on the consumer market, the other subgroup planned attacks on the corporate and institutional markets. As the game unfolded, the two subgroups pared down the number of plausible attacks. Eventually, the consumer subgroup decided to bombard automated clearinghouses with fraudulent (unauthorized), but legitimately formatted, transactions in something like a denial-of-service attack, while the corporate/institutional subgroup targeted backup systems, data repositories and money flows to and from the Federal Reserve system.

## Financial Services Makes a Likely Target for Cyberterrorists

Targeting the financial service industry would make sense for terrorists who wanted to achieve the objective on which the war game was based: undermine confidence in the United States as an economic power.

Successful attacks would do enormous damage to individuals and the financial system as a whole. In a post-event survey and evaluation, the war gamers were asked to look at their work and consider how devastating their scenario would be on the industry and, in their best judgment, how it would compare to a real attack. On a scale of one to 10, with 10 being the worst, the war game scenario was rated a 7.17. Financial service executives thought an actual attack would be 5.98 in devastation. The survey also revealed the war gamers believed a cyber attack was likely by mid-2004.

However, the complexity of the financial service industry makes it difficult, even for insiders, to determine what kind of attacks would cause enough damage to reach the goal of undermining confidence in the U.S. financial system. For example, the corporate/institutional subgroup, for a long time, considered an attack corrupting the systems of credit rating agencies, such as Moody's and Standard & Poor's, would not instantly wipe out a substantial amount of wealth.

## What Can Make Financial Service Providers More Secure

The financial group concluded that the mainframe roots of the technologies now used in its industry make its systems rather strong and defendable. For instance, legacy automated teller machines (ATMs) and teller systems often don't interface well. That not only reduces operating efficiency, but also makes it easier to contain a cyberattack to one system. Many new technologies promise improved performance and access, while failing to support minimal security infrastructure requirements.

The new technologies of greatest concern include:

- Web services
- Wireless
- Instant messaging
- Remote control technology
- Single sign-on
- Open-source Linux

Participants had the following recommendations for improving security:

- Do the basics. A well-designed infrastructure with firewalls and good internal security, as well as regularly patching vulnerabilities, will help significantly. Appoint someone internally as a "terrorist" and have the person infrequently attempt to break into your systems. Monitor workflow and reaction to the break-in attempts, and re-engineer policies and processes as appropriate.

• Take a global view of security. Financial service providers that have taken all necessary steps to secure their core operations in the United States could still be vulnerable there. For example, workers in a country with lower-security could mount a corrupt backup tape at a subsidiary with a direct link to the U.S. mainframe. Moreover, electronic links between other countries and the United States (even from subsidiary operations) could be subverted, thereby jeopardizing the integrity of the entire system.

• Ensure the physical security of all backup tapes, especially while in transit. Use cameras to monitor all facilities and machinery.

• Strengthen background checks of new hires, consultants, contractors and third-party administrators. Many cyberattacks proposed during the war game required infiltrating facilities.

• Review all insurance policies and consider the return on investment of clauses about the mitigation of cyberterrorism.

• Send out requests for information for pattern recognition software that can help rapidly detect aberrant activity (albeit after the fact).

• Avoid new technologies until they have been tested and found to include adequate security.

• Reconsider the security of disaster recovery and business continuity plans and operations. Even though the terrorists may be prevented from accessing "live" systems, they may still be able to corrupt backup information, thereby creating substantial inconvenience and potential losses in the event of recovery.

As part of their operational risk evaluations, rating agencies may well take into account a corporation's security. This judgment will influence market perception, and shareholders will not forgive failures to mitigate or possibly prevent the effects of this perception of elevated risk. If the terrorists don't get you, the market will.

# 4. THE INTERNET

The Internet group created feasible plans to control or disable the Internet. However, the group found that following security best practices can dramatically mitigate the threat from external cyberattacks.

## How the Game Played Out

The Internet group identified four goals and organized into groups to develop plans for accomplishing them:

- Reconnaissance to identify vulnerable Internet sites
- Seizing control of the Internet
- Disrupting operations on the Internet
- Disabling the Internet

Seizing control required a high level of technical expertise. With the right technical expertise, the goal seemed possible. The group developed a series of plans capable of achieving the four goals by using Internet technologies such as viruses, worms and other malicious code delivered by subverting or releasing peer-to-peer (P2P) applications, spyware and other freeware.

Because elements of the Internet consist of relatively new technologies that enterprises have adopted only in recent years, judging business dependencies remains difficult. Therefore, assessing how much damage widespread cyberattacks could cause remains more difficult for this industry than for more-established industries.

## Internet Terrorism Will Look Like Business as Usual

Of the four industry groups — telecommunications, Internet, electrical power grid, financial services — involved in the war game, the Internet group's plans seemed most difficult to uncover. It relied less on physical attacks or malicious insiders placing moles in enterprises and more on deception through the use of nearly "transparent" means in the ordinary use of the Internet. A few people with minimal outside contact would suffice. Thus, Internet terrorism would likely be indistinguishable from legitimate operations or "normal" cyberattacks — or at least won't look any different from commonplace mischief until it is too late for enterprises to counter the attack meaningfully.

For example, using normal channels for software delivery, insiders can distribute illegitimate or undesired code — much as happened with Kazaa. Kazaa's software contained code that tracked user behavior, although users did not know this when they downloaded the software. However, tight quality assurance processes can limit such attempts.

## What to Do

To avoid being used for, or targeted by, an Internet cyberattack, enterprises should follow security and business continuity best practices:

• Implement layered security defences for desktops, servers and networks, including such tools as properly configured firewalls, intrusion prevention or detection, antivirus and strong authentication.

• Evaluate new patches and updates on release, and develop a rating method and process for implementing patches [1], [2].

• Maintain a proactive overall security posture. Monitor vulnerability and alert distribution lists, as well as internal logs, for possible security incidents.

• Test physical security procedures, including access to facilities, background investigations on employees and interaction with electronic systems.

• Ensure that critical decision makers and the computer-incident response and crisis-management teams can communicate in several ways and that they understand crisis-management procedures.

• Update virus signatures daily, if not more frequently.

• Perform vulnerability assessments, including penetration testing at least annually, using a third party.

• Manage user accounts. Disable inactive accounts, review password reset procedures, limit administrative access and rotate passwords, remove default accounts and consider strong authentication.

• Constantly monitor publicly accessible Web sites for possible security breaches.

• Examine security practices for remote access, including dial-up lines, extranets and virtual private networks.

• Review the security policies of external service providers and other business partners, and include security service-level agreements in new contracts.

## Even More Basic Issues

According to Gartner research, which the gamers confirmed, many enterprises do not regularly follow all of these best practices. The primary reason is that many are reluctant to sacrifice a little efficiency or convenience for the sake of better security. Thus, the most fundamental way to protect the enterprise from cyberattacks is to create a security-conscious culture within the enterprise.

Accordingly, Gartner and the war game participants recommend the following:

• CEOs, CIOs and CFOs should set the tone for their enterprises: Make Internet security as high a priority as serving the application, infrastructure and storage requirements of internal and external customers.

• Use the enterprise's buying power to give preference to software with higher quality and security.

• CEOs and CIOs should insist that employees follow security best practices and should mandate periodic audits. They should reward managers who demand good security practices from employees at all levels. Most security best practices aren't followed because they are inconvenient — for example, users often complain that changing passwords regularly is disruptive. A good IS organization treats employees as customers and tends to go along with their preferences, and CEOs do, too. Both groups dislike hearing people complain about something as mundane as password updates.

• Prevent insecure software from entering the enterprise by enforcing policies that prevent employees from casually downloading code, especially P2P technologies and spyware.

• Promote higher-quality control procedures when creating programming code, including testing for vulnerabilities, as well as for "bugs."

• Include wide-scale cyberattacks in business continuity plans, and consider partnering with a peer enterprise for mutual aid.

# 5. TELECOMMUNICATIONS

Even the attack planners in the telecommunications group thought that the possibility of terrorists bringing down the whole system was remote. Unlike the Internet, which consists largely of common protocols and code running on top of network infrastructure, the telecom network contains many distinct areas, each with intricate technologies that would make it much more difficult for a terrorist group to master. However, individual vulnerabilities remain. More importantly, the industry relies so heavily on engineering good security that it sometimes neglects good security practices and discounts the need to react quickly to emerging threats.

## How the Game Played Out

The group broke the telecom system into six or nine constituent parts, from signaling to operations support system (OSS), and designed attacks to destroy each part. The telecom group felt that if terrorists could accomplish the tasks it outlined, the whole telecom network — that is, the ability of Americans to communicate in real time by voice and data — would shut down for a long period.

## Why the Telecom System Is Secure

The scenario sounds a lot more frightening than it really is. At most, terrorists could probably attack one or two parts. Such attacks would cause some damage but would likely not represent a serious, long-term setback for the nation as a whole.

Why wouldn't something worse happen? Again, the technology involved in each area is so complex, it takes an entire career to accumulate the expertise to do real harm to any one part. Even seasoned industry veterans don't know much about areas outside the one they work in. Thus, it would be very difficult for a terrorist group to accumulate enough knowledge in enough areas to threaten the telecom system as a whole.

In addition, unlike the software industry, which has traditionally produced code very quickly with lax security, the telecom industry has always known that security is fundamental to the products and services it provides. Systems have been engineered with failsafes and ample redundancies so that it would be very difficult, for example, to take over a network switch the way a good hacker could take over an Internet-connected server.

## Where Vulnerabilities Remain

This excellent engineering should not lure the telecom industry into a false sense of security. During the war game, the telecom group identified several vulnerabilities that should be closed, and Gartner and the U.S. Naval War College will pass that information on to the responsible parties.

The industry's attitude toward security should change as well. For example, the Federal Communications Commission (FCC) does not mandate that carriers use best practices. That approach worked when AT&T held a telecom monopoly and understood its responsibility. In a deregulated market, several large, and many small, competitive carriers operate, and not all of them follow security best practices. In addition, enterprises often outsource their networking or deal with carriers that use a second carrier's networks to deliver service. Although the enterprise and primary carrier may follow best practices religiously, the secondary providers may not; that situation opens a possible channel for terrorists to attack even the scrupulous enterprise or primary carrier.

The telecom industry tends to react slowly when vulnerabilities are discovered in networking protocols. Often, it takes a year for industry committees to come up with a patch. The sloppy IT industry could teach the telecom industry a lesson here. The IT industry knows it does a poor job of engineering security with software; therefore, it responds with great urgency when a vulnerability comes to light. Patches appear in weeks or days. A slower response increases the opportunity for terrorists.

The telecom system remains very secure, but terrorists often surprise planners in what they can accomplish. Accordingly, Gartner recommends the following to participants in the industry:

- The FCC should prod the industry to come together and ensure it follows best practices to guard against such attacks. The FCC should avail itself of the expertise and recommendations of the National Security Telecommunications Advisory Council, a committee formed 20 years ago as a means for telecom companies and government to work together to advance the security and reliability of the public telecommunications infrastructure.

• Carriers should develop a way to deal with emerging vulnerabilities as urgently as the IT industry does.

• Vendors should create tools for maintaining secure systems. For instance, the IT industry has scanning and intrusion detection tools for its networks; the telecom industry should try to develop similar tools for its more-complex environment.

• Enterprises should ensure that their service providers have done what is necessary for the physical and cybersecurity of their systems.

# 6. SUMMARY AND RECOMMENDATIONS

The cyberterrorism threat is not a technical issue. Terrorists above all would want to show the public that a devastating attack had occurred. The targets have to be highly visible, and the attacks have to destroy the morale of the average person. Attacking the telecom system couldn't simply damage equipment but would have to undermine people's ability to communicate in dealing with the crisis. Terrorists wouldn't inflict the most damage by taking down the insurance industry because people would still live in their houses even if they couldn't get a homeowner's policy; equity markets offer a better target. The Internet group thought it could shut down networks completely when it wanted. Instead, it crafted attacks to use the Internet to distribute false or inaccurate information — say, about credit card transactions — to drive business away from the Internet. Therefore, enterprises should think of efforts to combat cyberterrorism as contributing to their strategic goal of retaining customer confidence.

## Plan on Several Fronts

Terrorists will try to find the most effective combination of several different types of attacks, including cyberattacks via the Internet, installing devices with corrupted software at the core of the infrastructure (such as to control the power grid) and physical attacks on key facilities. In addition, a digital Pearl Harbor would most likely combine attacks on several different infrastructure areas. In response, enterprises must do more than find and plug security holes. They should assume a more comprehensive perspective. That means viewing even minor problems as potentially contributing to a wider attack. A few stolen credit card numbers or botched transactions may do little more than inconvenience an enterprise, but more occurrences at other companies could contribute to a serious assault on the nation's financial system. Thus, Gartner recommends that each enterprise create a management function to coordinate analysis of the various stresses affecting it, whether an obvious cyberattack or a seemingly unrelated anomaly such as the theft of an employee's laptop (what information is on the laptop?).

In addition, the enterprise should communicate with other companies about these threats to try to detect larger patterns. The United States may need to require the reporting of IT anomalies in certain key critical industries. For example, the financial service group suggested increasing sensitivity requirements and shortening the deadline for reporting anomalies discovered in the processing of transactions. It also felt a central clearinghouse for such reports from all financial institutions could help identify potentially crippling attacks early on.

A digital Pearl Harbor, even though unlikely, represents a real threat to the United States or any other country whose economy depends upon the IT infrastructure functioning uninterruptedly. And though a DPH may be unlikely, there are many lesser scenarios encompassed within DPH that are simpler to carry out and would nonetheless create havoc. The critical shortage of IT staff and the lax screening of employees make it impossible to discount the possibility that an attack may already be underway. However, this threat may not be easy to detect because dedicated terrorists will try to break their attack into small functions that look relatively harmless in isolation and will attempt to exploit compartmentalized business processes and regulatory requirements. Each enterprise should look out for signs of a potentially crippling blow to itself and to the nation.

## Keep an Open Mind About Threats

The Sept. 11 attacks suggested that a dedicated group with small resources could perform the intricate choreography necessary for a devastating cyberattack on the IT infrastructure of the United States. Not surprisingly, most participants seemed to imagine the attackers as al Qaeda members. However, such an assumption could be fatal. An employee in any key industry could do more damage from the inside than al Qaeda could do from the outside. Or a terrorist group could offer money to induce an employee to play a key role in a much larger plot that he knows nothing about — all he knows is that he's getting paid well to introduce a bit of code into a router. Attackers wouldn't have to activate such a Trojan Horse or backdoor for three years or more.

Terrorists have many options, and they will always look for where enterprises are most vulnerable and attack them there. They also have the weapon of patience. Mohammed Atta started planning six years before Sept. 11, a longer period than this war game allowed participants. Follow the advice of Richard Clarke. At the time of the DPH game, he was the U.S. government's top IT security official and the man who coined the phrase "digital Pearl Harbor". He tells companies: "Don't try to guess what the threats are. Find out what vulnerabilities you have that any terrorist could exploit, then fix them."

## Conclusions

This vulnerability stems from the generally poor quality of software sold today and from the availability of information on the Internet that terrorists need. Because enterprises own and operate so much of cyberspace, they will have to take an active part in defending against cyberterrorism. Enterprises that contribute most to national security should understand:

- The real target in any cyberattack by terrorists will likely be the general loss of confidence in the system under attack as well as its functions.

- Who will attack what targets at what time and with what means cannot be determined ahead of time with great precision.

- Addressing the terrorist threat requires a comprehensive and collaborative approach.

## REFERENCES

1. **John Pescatore - "Internet Vulnerability Risk Rating Methodology"** http://www.gartner.com/DisplayDocument?doc_cd=102814
2. **Dean Lombardo- Patch Security Holes but Demand Better Security From Vendors** http://www.gartner.com/resources/103300/103382/103382.pdf

### About the Authors

Eric Purchase has written for IT industry research firm Gartner, Inc. since 1997. An independent scholar, he is author of Out of Nowhere: Disaster and Tourism in the White Mountains (Johns Hopkins University Press, 1999), which shows how writers, painters, scientists and entrepreneurs helped create the tourist industry in New Hampshire's White Mountains. He holds a Ph.D. in Comparative Literature from the University of Connecticut.

French Caldwell is a vice president in Gartner Research, where he directed the Digital Pearl Harbor project. His research includes analysis of knowledge management services and technologies, public technology policy and homeland security strategies. Prior to joining Gartner, Mr. Caldwell was the director of knowledge services in a global consulting practice where he worked with strategic clients, including the Central Intelligence Agency and the Department of Defense. Mr. Caldwell completed a career as a nuclear submarine officer, and has directed special congressional projects for the Secretary of the Navy and the Secretary of Defense. He is an adjunct fellow at the Center for Strategic and International Studies and a former federal executive fellow at the prestigious Brookings Institution. Mr. Caldwell holds a bachelor of science degree in oceanography from the United States Naval Academy and a master's degree in international studies from Old Dominion University. He attended the U.S. Navy Nuclear Engineering Program and the U.S. Naval War College.

Chapter 5

# CYBERCRIME AND THE LAW

*Balancing Privacy Against the Need for Security*

Susan W. Brenner
*University of Dayton School of Law*

**Abstract**: The proliferation of cybercrime, combined with the tragic events of September 11, 2001, has placed a premium on security, both from crime and from terrorism. At the same time, it has raised important issues concerning privacy.

**Keywords**: cybercrime

The proliferation of cybercrime, combined with the tragic events of September 11, 2001, has placed a premium on security, both from crime and from terrorism. One aspect of this emphasis on security has been the adoption of new legal measures which give law enforcement greater authority to investigate and prosecute cybercrime and cyberterrorism, among other evils. But while most Americans support efforts to prevent future acts of terrorism and to minimize our victimization by cybercrime, the breadth of the authority given law enforcement, and especially federal law enforcement, has moved some to express concern that our fundamental right to privacy is being eroded by these measures. This chapter examines the legal foundations of our right to privacy and the post-9-11 legislation that was enacted to facilitate the war on terrorism and the battle against cybercrime. It analyzes the tension that exists between the two and considers how the future balance between privacy and security should be struck.

## 1. PRIVACY GUARANTEES

In the United States, citizens' fundamental rights derive from the Constitution, which is the foundation of the American legal system. The

Constitution itself defines the system of government that exists at the federal and state levels; it does not address the issue of individual rights. This issue was addressed separately in the Bill of Rights, the first ten amendments to the Constitution.

The Constitution itself was drafted and then opened for ratification by the states in 1787; after being ratified by the necessary number of states, it went into effect on March 4, 1789. As noted above, the Constitution articulates a set of general principles which dictate the structure of the government that was to be instituted, including the separation of power among three branches, legislative, executive and judicial. It does not speak to the issues of individual rights or checks upon the power of that government. Many of the representatives to the Constitutional Convention expressed concern about this lack; they wanted clauses included that explicitly declared the rights belonging to individual citizens and that specifically restricted the government's power in various areas. Instead of incorporating these provisions into the Constitution itself, they were added in the form of ten amendments, which have since become known as the Bill of Rights [1]. The Bill of Rights was quickly drafted, ratified by the necessary number of states and went into effect on December 15, 1791 [2].

Neither the Constitution nor the Bill of Rights refers to a "right to privacy." Indeed, the word "privacy" does not appear in either document. But in the slightly more than a century since the Bill of Rights was adopted, the U.S. Supreme Court has held that a general right to privacy derives from several of the amendments which compose the Bill of Rights. The Court has held that the First Amendment, which establishes the rights to free speech and a free press, creates the right to speak anonymously, i.e., to preserve one's identity as private when speaking out on religious, political and related issues [3]. It has also held that the Fourth Amendment, which creates a right to be free from "unreasonable" searches and seizures, creates a right to privacy in "protected" areas such as one's home or business and in one's person [4]. This does not mean police cannot search a place or a person; it means they cannot do so unless they have a valid search warrant or can invoke an applicable exception to the warrant requirement. And, finally, the Court has recognized that the Fifth Amendment essentially creates a right to privacy in one's thoughts; that is, the Fifth Amendment prohibits the government from "compelling" anyone to testify against himself or herself if the testimony can be used to prosecute that person for a crime [5]

At the time the Constitution and the Bill of Rights were adopted, there was little, if any, technology. Consequently, these Constitutional guarantees were all crafted in the context of law enforcement activities that were confined to the real-world, such as forced entries to search homes and businesses. The Supreme Court therefore was not called upon to consider how, if at all, technology impacts upon these guarantees until the first part of the twentieth century, when it had to decide if telephone communications were protected by the Fourth Amendment. The issue in *Olmstead v. United States*, 277 U.S. 438 (1928), was whether federal agents violated the Fourth Amendment when they installed wiretaps on telephone lines and used them to listen in to conversations Roy Olmstead, the suspected leader of a large bootlegging operation, had with several of his associates. Olmstead and the people with whom he spoke were in their homes or businesses when they had these conversations; it was clear that the Fourth Amendment would have prohibited the agents from entering these premises and eavesdropping on what was said. It was not clear if using technology to intercept the conversations as they passed over the telephone wires violated the Fourth Amendment; this was the Court's first venture into deciding how technology impacts on the right to privacy. In an opinion written by Chief Justice Taft, a majority of the Court simply transposed traditional Fourth Amendment standards into this new context, holding that the agents did not violate the Fourth Amendment because they did not physically enter into a constitutionally protected area. Justice Brandeis dissented, arguing that "'in the application of a Constitution, our contemplation cannot be only of what has been, but of what may be'" and maintaining that what we would now call a "technologically-neutral" interpretation of the Constitution was necessary to preserve the values it was intended to protect. In a famous passage, Justice Brandeis observed that when the

> Fourth and Fifth Amendments were adopted . . . . [f]orce and violence were then the only means . . . by which a government could directly effect self-incrimination. It could compel the individual to testify . . . by torture. It could secure possession of his papers and other articles incident to his private life . . . by breaking and entry. . . . But. . . . more far-reaching means of invading privacy have become available to the government. . . .
>
> The progress of science in furnishing the government with means of espionage is not likely to stop with wire tapping. Ways may some day be developed by which the government, without removing papers from secret drawers, can reproduce them in court, and . . . expose to a jury the most intimate occurrences of the home. . . . Can it be that the Constitution affords no protection against such invasions of individual security?[6]

It was not until 1967, in *Katz v. United States*, 389 U.S. 347 (1967), that the Supreme Court finally recognized that the literal application of standards developed to deal with the intrusions encountered in centuries past was not a viable way to preserve and enforce privacy guarantees in an age of electronic communication. The issue in the case was was whether the government had violated Katz's Fourth Amendment rights by using a wiretapping device installed on the outside of a phone booth to record calls he made while inside. The *Katz* Court rejected *Olmstead* in favor of the approach Justice Brandeis had advocated forty-years before:

> [A] person in a telephone booth may rely upon the protection of the Fourth Amendment. One, who occupies it, shuts the door behind him, and pays the toll that permits him to place a call is surely entitled to assume that the words he utters into the mouthpiece will not be broadcast to the world. To read the Constitution more narrowly is to ignore the vital role that the public telephone has come to play in private communication [7]

In 1968, Congress responded to the *Katz* decision *Berger v. New York,* 388 U.S. 41 (1967), which also dealt with wiretapping, by enacting legislation – known as "Title III" – that implemented procedural protections for oral and wire communications [8]. Title III generally prohibited the interception of such communications except in accordance with procedures set out in Title III; these procedures satisfied the Fourth Amendment since they required law enforcement to show probable cause and obtain what was in effect a warrant before interception telephonic and other wire communications [9]. Title III was adequate for wire communications, but the next decade or so saw the emergence of new technologies which, in turn, resulted in new modes of communication. In 1986, Congress responded by adopting the Electronic Communications Privacy Act or ECPA. ECPA amended Title III to protect electronic communication such as e-mail, and added Title II, which protects stored wire and electronic communications [10].

## 2. USA PATRIOT ACT

The USA Patriot Act [11] was a direct response to the terrorist attacks of September 11, 2001. It was enacted six weeks after the attacks; President Bush signed the bill on October 26, 2001. The Patriot Act is 342 pages long and contains ten titles. Among other things, it does the following: expands the federal definition of "terrorism" to include domestic as well as international terrorism; gives law enforcement officers greater surveillance and investigative authority for domestic law enforcement and for foreign intelligence gathering; strengthens laws against money laundering and increases the obligations imposed on financial and other institutions to monitor and report suspected money laundering; and creates new crimes plus increases penalties for existing crimes. An exhaustive treatment of the provisions of the Patriot Act are beyond the scope of this chapter; it merely provides an overview of the changes the Patriot Act made in federal law [12]

One change the Patriot Act made was to Title III. Prior to this revision, Title III wiretaps, which are discussed in the prior section, could not be used in cybercrime investigations -- i.e., investigations of conduct violating 18 U.S. Code § 1030 -- because section 1030 was not a Title III predicate offense [13]. The Patriot Act remedied this, adding section 1030 to the Title III list of predicate offenses, which means wiretapping can now be used in federal cybercrime investigations [14]

The Patriot Act also contained amendments directed at the substance of 18 U.S. Code § 1030, which is the general federal cybercrime provision, i.e., the statute that defines the federal cybercrime offenses. These amendments did the following:

(1) raised the maximum penalties for hackers who damage computers protected by the statute to ten years for first offenders and twenty years for repeat offenders;

(2) made it clear that an individual violates the statute if he or she intends to cause damage to a computer, computer system or the information contained thereon – the government is not required to prove that the hacker intended any specific damage or to cause any specific amount of damage;

(3) allow the government to aggregate damage inflicted by a hacker within a one-year period to meet the $5,000 loss threshold required to establish jurisdiction over certain of the offenses defined by the statute;

(4) make it an offense for a hacker to damage a computer used by or a for a government entity in the administration of justice, national security or national defense even if the damage does not result in provable loss exceeding $5,000;

(5) expand the scope of the statute so that it now applies to actions taken against computers located outside the United States as well as those within the United States – computers located abroad come under the protection of the statute if their use affects interstate or foreign commerce or "communication of the United States";

(6) allow a court to consider state hacking convictions in imposing sentence for violations of the federal statute; and

(7) defines the "loss" needed to trigger application of the statute as including "a wide range of harms typically suffered by the victims of

computer crimes, including costs of responding to the offense, conducting a damage assessment, restoring the system and data to their condition prior to the offense, and any lost revenue or costs incurred because of interruption of service." [15]

In addition to making these changes to the substance of 18 U.S. Code § 1030, section 308 of the Patriot Act incorporated two provisions of section 1030 into the definition of the federal crime of terrorism which is contained in 18 U.S. Code § 2332b. Originally, the Act would have incorporated all of the section 1030 offenses, which would have meant that minor offenses such as web site defacement and intrusions not causing damage would have been defined as a federal crime of terrorism [16]. The Patriot Act does not go this far, but it does bring the following within the statutory definition of a federal crime of terrorism when the conduct is calculated to influence or affect the conduct of government by intimidation or coercion or to retaliate against government conduct:

(1) violating 18 U.S. Code § 1030(a)(1) by accessing restricted or classified information on computers that require protection for reasons of national security or national defense and, having reason to believe the information could be used to injure the United States or to the advantage of a foreign nation, communicates that information to one who is not authorized to have it; and

(2) violating 18 U.S. Code § 1030(a)(5)(A)(i) and thereby causing damage that results in physical injury, danger to public heath or safety or the impairment of medical care or that affects a computer system which is used by or for a government entity in furtherance of the administration of justice, national defense, or national security [17]

There are several consequences of having conduct encompassed by 18 U.S. Code § 1030 fall within the definition of a federal crime of terrorism. These include, for example, a longer statute of limitations, alternate maximum penalties, the seizure of assets prior to conviction and the forfeiture of assets upon conviction and the possibility that the conduct can also be prosecuted as, among other things, providing support to or harboring terrorists [18]

The Patriot Act also made other changes that enhance investigators' ability to gather evidence in cybercrime investigations, as well as in investigations of terrorism and other crimes. One of these changes lets authorities use pen register and trap and trace devices to "capture source and addressee information" for e-mail [19]; they cannot, however, use this authority to intercept the content or the subject header of e-mails [20]. The ability to use pen register and trap and trace devices for e-mail is an advantage because the burden of obtaining authorization for this type of surveillance is much less than that which applies under either Title III, the Electronic Communications Privacy Act or the Fourth Amendment [21]. Traditionally, pen register and trap and trace devices were used to identify the "source and destination of calls made to and from a particular telephone."[22]. They issue upon a court order approving their use [23] These orders "are available based on the government's certification, rather than a finding of the court, that the use of the device is likely to produce information relevant to the investigation of a crime, any crime" [24]. The "relevant" standard is a lower evidentiary standard than the probable cause required for other types of surveillance. If the investigating agents use the government's own technology (such as the program formerly known as Carnivore), they must, within thirty days of the surveillance, provide the following information to the court under seal: "(1) the identity of the officers who installed or accessed the device; (2) the date and time the device was installed, accessed, and uninstalled; (3) the configuration of the device at installation and any modifications to that configuration; and (4) the information collected by the device. 18 U.S.C. § 3123(a)(3)."[25]

Other sections of the Patriot Act amend the Electronic Communications Privacy Act (ECPA), making it easier for agents to obtain access to voice mail and to obtain user information from Internet Service Providers. As to the former, the revisions let agents use the ECPA procedures instead of the more demanding Title III procedures to obtain access to stored voice mail [26]. Under prior law, agents had to use "a wiretap order to obtain voice communications stored with a third party provider but could use a search warrant if that same information were stored on an answering machine inside a criminal's home."[27]. It was the Department of Justice's view that stored voice mail did not require the heightened privacy protections given to the interception of real-time voice communications: "Stored voice communications possess few of the sensitivities associated with the real-time interception of telephones, making the extremely burdensome process of obtaining a wiretap order unreasonable" [28]. As to obtaining evidence from Internet Service Providers, provisions of the Patriot Act expanded the amount of information agents can obtain with a subpoena. Under prior law, they could use a subpoena to compel an ISP to provide them with "a limited class of information, such as the customer's name, address, length of service, and means of payment."[29]. They could not, however, use a subpoena to obtain credit card numbers or other information that could be used to establish a customer's identity; according to the Department of Justice, this information was important because many users, especially those bent on committing crimes, use false names when establishing accounts with an ISP

[30]. As a result, the Patriot Act expanded the scope of the information agents can obtain with a subpoena. The Patriot Act amendments let them compel an ISP to provide them with the following: (a) records of session times and durations; (b) temporarily assigned network addresses; and (c) the means and source of payment a customer uses to pay for his account with an ISP, including credit card numbers or bank account numbers [31].

The Patriot Act also gave ISP's the ability to share information with investigating authorities on their own initiative. This ability takes two forms, one of which involves the safety of others, the other of which involves the ISP's own security. As to the former, the Patriot Act amendments expand the scope of the information an ISP can provide to the authorities in an emergency, i.e., if the service provider learns that a customer is part of a conspiracy to commit an imminent terrorist attack [32]. Prior law did not permit an ISP to disclose customer records or communications in a situation such as this. The law as amended by the Patriot Act permits, but does not require "a service provider to disclose to law enforcement either content or non-content customer records in emergencies involving an immediate risk of death or serious physical injury to any person."[33]

As to ISP security, the Patriot Act made two changes in the law to enhance an ISP's ability to work with authorities to protect itself from attacks. The first change altered a peculiar inconsistency in existing law: "[P]rior to the Act, the law did not expressly permit a provider to voluntarily disclose non-content records (such as a subscriber's login records) to law enforcement for purposes of self-protection, even though providers could disclose the content of communications for this reason."[34]. The Patriot Act amendments eliminated this inconsistency and made it clear that ISP's "have the statutory authority to disclose non-content records to protect their rights and property."[35]. The other change eliminated an ambiguity in the law governing an ISP's ability to obtain law enforcement assistance in monitoring trespassers on their systems:

> Although the wiretap statute allows computer owners to monitor the activity on their machines to protect their rights and property, until . . . the [Patriot] Act was enacted it was unclear whether computer owners could obtain the assistance of law enforcement in conducting such monitoring. This lack of clarity prevented law enforcement from assisting victims to take the natural and reasonable steps in their own defense that would be entirely legal in the physical world. In the physical world, burglary victims may invite the police into their homes to help them catch burglars in the act of committing their crimes. The wiretap statute

> should not block investigators from responding to similar requests in the computer context simply because the means of committing the burglary happen to fall within the definition of a `wire or electronic communication' according to the wiretap statute. Indeed, because providers often lack the expertise, equipment, or financial resources required to monitor attacks themselves, they commonly have no effective way to exercise their rights to protect themselves from unauthorized attackers [36].

To remedy this deficiency, the Patriot Act, amendments specifically allow victims of computer attacks to authorize law enforcement agents to monitor trespassers on their systems. But before law enforcement monitoring can occur, four requirements must be met: (1) the owner of the computer/computer system must authorize the interception; (2) the person conducting the interception must be lawfully engaged in an investigation, either a criminal investigation or a foreign intelligence investigation; (3) the person conducting the interception must have "reasonable grounds" to believe that the contents of the communications to be intercepted are relevant to the investigation and must intercept only communications sent or received by the trespasser; and (4) "computer trespasser" is limited to someone accessing a computer/system without authorization and does not include anyone who has a contractual relationship with the owner of the system which allows the individual access to that system [37]

The Patriot Act made a number of other changes in an effort to improve law enforcement's ability to investigate terrorism and other crimes, including cybercrimes. Only two of those additional changes are discussed here; for a detailed analysis of what the Patriot Act did, the reader is encouraged to consult other sources [38]

The first change concerns foreign intelligence gathering pursuant to the Foreign Intelligence Surveillance Act (FISA) and its relationship to the investigation of domestic crime. It is perhaps important to underline the distinctions between the two: Domestic criminal investigations seek evidence about, and persons believed to be involved in, the commission and/or contemplated commission of acts that are unlawful under federal and state criminal law. Foreign intelligence investigations seek evidence concerning other countries and their citizens; these investigations are not "limited to criminal, hostile, or even governmental activity. Simply being foreign is enough."[39]

Federal law restricts foreign intelligence gathering within the United States because of citizen concerns about the creation of a domestic secret police, concerns that were exacerbated by certain law enforcement activities carried out during the Vietnam War era [40]. The restrictions are contained in FISA:

> Congress enacted the Foreign Intelligence Surveillance Act (FISA), something of a Title III for foreign intelligence wiretapping conducted in this country, after the Supreme Court made it clear that the President's authority to see to national security was insufficient to excuse warrantless wiretapping of suspected terrorists who had no identifiable foreign connections, *United States v. United States District Court*, 407 U.S. 297 (1972). FISA later grew to include procedures for physical searches in foreign intelligence cases, . . . for pen register and trap and trace orders, . . . and for access to records from businesses engaged in car rentals, motel accommodations, and storage lockers [41]

The Patriot Act made a number of modifications in the FISA procedures; these modifications were designed to ease the restrictions that had been placed on foreign intelligence gathering in the United States and to give foreign intelligence investigators greater access to evidence obtained in the course of domestic criminal investigations [42]. Perhaps the most significant of these modifications concerns the showing agents have to make in order to obtain a court order authorizing searches or surveillance. Prior to the Patriot Act, agents could obtain these orders, which are issued in secret by a special FISA court, only when foreign intelligence gathering was *the* reason for a search or surveillance. The Patriot Act relaxed this standard: Agents can now obtain these orders if foreign intelligence gathering is *a significant* reason for a search or surveillance [43]. This modification has caused great concern among privacy advocates, who contend that it gives the government "a green light to resume spying on government enemies."[44]

The Patriot Act also made certain modifications in the procedural rules that govern the conduct of criminal – versus foreign intelligence – investigations. Two of those modifications are significant enough to warrant discussion here.

The first modification concerns the conduct of grand jury proceedings, which are encompassed by a rule of secrecy [45]. In the federal system, grand jury secrecy is imposed by Rule 6(e) of the Federal Rules of Criminal Procedure [46 Rule 6(e) declares that except for witnesses who testify before a grand jury, all others having access to "matters occurring before a grand jury" must maintain the secrecy of that information or face prosecution for criminal contempt and/or related offenses [47]. The rule of secrecy is bound up with the purpose of grand juries, which is to investigate to determine if a crime has been committed and return an indictment, a set of charges, if it finds probable cause to believe a crime has been committed [48]. Because grand juries sit to investigate criminal activity and because their activities are shrouded in secrecy, grand juries could not, until the adoption of the Patriot Act, share evidence they discovered with persona outside the criminal justice system, such as foreign intelligence investigators, those charged with maintaining national security and the like [49].

The Patriot Act altered this: When it was being drafted, the Department of Justice recommended that it alter Rule 6(e) so that the rule would allow grand jury material – evidence, transcripts of witness testimony and other matter – to be shared with federal law enforcement personnel not involved in that grand jury's investigation, national defense and national security personnel and immigration officers whenever the material pertained to national security or terrorism [50]. The recommendation was adopted and section 203(a) of the Patriot Act modifies Rule 6(e) to allow matters occurring before a federal grand jury to be disclosed to "any federal law enforcement, intelligence, protective, immigration, national defense, or national security' officer to assist in the performance of his official duties" [51]

Those who opposed this modification in Rule 6(e) argue that it could result in the grand jury's being abused, i.e., being used for "intelligence gathering purposes, or less euphemistically, to spy on Americans."[52].

The other general procedural modification the Patriot Act made involves search warrants. The process for obtaining and executing a search warrant is dictated by Rule 41 of the Federal Rules of Criminal Procedure. Under pre-Patriot Act procedure, agents who obtained a search warrant went to the place to be searched, gave the person who owned or was in charge of the premises a copy of the warrant, conducted the search, seized evidence that was within the scope of the warrant, and then gave this person an inventory of the items they were taking away with them, as having been seized pursuant to the warrant [53] When the Patriot Act was being drafted, the Department of Justice recommended that it authorize the issuance of "sneak and peek warrants," a practice that appeared in the 1990's and had been upheld by some federal courts [54]. A "sneak and peak warrant" authorizes agents "to secretly enter, either physically or virtually; conduct a search, observe, take measurements, conduct examinations, smell, take pictures, copy documents, download or transmit computer files, and the like; and depart without taking any tangible evidence or leaving notice of their presence."[55]. In 1999, for example, federal agents used a sneak and peek warrant to break into the office of alleged Mafia loan shark Nicodemo Scarfo and install a keystroke logger on his computer [56].

Despite opposition from privacy advocates, section 213 of the Patriot Act approves the use of sneak and peek warrants. As one author explained, sneak and peek warrants are a "radical departure from conventional search warrants" for several reasons [57]. For one thing, they authorize police to "effect physical entry into private premises without the owner's or the occupant's permission or knowledge to conduct a search" [58]; generally, as in the Scarfo case, this entry requires police to break and enter the premises. Also, when police use a sneak and peek warrant, the search occurs only when the occupants are absent from the premises. The entry, search, and any seizures are conducted in such a way as to keep them secret. The search and seizure focus on obtaining intangibles, i.e., information concerning what has been going on, or now is inside, the premises. Photographs may be taken. Usually, no physical objects are removed. If objects are removed this is accomplished in such a way that the removal remains clandestine; for example, an item seized might be replaced with another item that appears to be the original. No copy of the warrant or receipt is left on the premises. Sometimes the same premises is subjected to repeated covert entries under successive warrants. Generally, it will not be until after the police make an arrest or return with a conventional search warrant that the existence of any covert entries is disclosed. This may be weeks or even months after the surreptitious search or searches [59].

The Patriot Act authorized sneak and peek warrants by amending 18 U.S. Code § 3103a, which deals with search warrants, to include a provision that allows courts to delay notifying someone that their property has been the object of a search (and a seizure of their property) for a "reasonable period" after the search (and seizure) has occurred. The amendment allows sneak and peek warrants to issue when the court finds there is "reasonable" cause to believe that notifying the person of the search (and seizure) would have an "adverse result" [60]. An adverse result is defined as endangering the life or safety of a person or creating a risk that a guilty person will flee prosecution, that evidence will be destroyed or altered or that witnesses will be intimidated; it also encompasses other consequences that would jeopardize an investigation or trial [61]. The Patriot Act amendment also creates a default preference for issuing sneak and peek warrants only when tangible evidence is not to be seized, but the court can authorize the use of a sneak and peek warrant to seize tangible evidence if it finds "reasonable necessity for the seizure" [62].

Privacy advocates are concerned that this acceptance of sneak and peek warrants will reinstitute the domestic spying practices and abuses that occurred during the Vietnam War era, especially since the use of sneak and peek warrants is not limited to terrorism investigations [63]. They can be used in an investigation of any federal crime, and it is highly likely that states will follow the federal government's lead and institute their own rules authorizing surreptitious entries.

## 3. SECURITY VERSUS PRIVACY

The adoption of the Patriot Act only underscores what was already apparent, i.e., that there is a fundamental tension between maintaining security (especially in an age of terrorism) and preserving individual privacy. The balance between privacy and security is an issue courts have struggled with for the more than a century that has elapsed since the ratification of the Bill of Rights; this struggle involves judicial and legislative efforts to ensure that law enforcement officers had the tools they needed to preserve domestic peace and order but at the same time guarantee that individual's rights to privacy were not violated.

Until very recently, this struggle was concerned only with "real world" privacy, i.e., with establishing boundaries for the measures law enforcement officers could take in the external, physical world. Courts articulated, for example, rules prescribing when law enforcement officers could, and could not, enter someone's home without a warrant or when and under what circumstances they monitor conversations occurring in the home or someone's activities outside the home. By the 1990's, these rules were, for the most part, well-established, so that it was reasonably clear what law enforcement could and could not do.

Two developments have raised new questions about how the balance between law enforcement authority and individual privacy is to be struck. One was the rise of computer technology and the migration of human activities into cyberspace; cyberspace is not the "real world," for which existing privacy guarantees were fashioned. This means that courts must grapple with a host of new issues. For example, the Supreme Court long ago held that the contents of a sealed letter are private, even while the letter is in transit via the postal service and cannot be accessed by law enforcement unless the officers first obtain a search warrant. Courts are now dealing with how, indeed, with whether, this rule should also apply to email. Some argue that email should be treated like a sealed letter and given constitutional protections of privacy. Others argue that email is like a post card; since the contents of a post card can be read by anyone who has occasion legitimately to come in contact with it, courts have held that the contents of a post card, unlike those of a sealed letter, are not private and are therefore not entitled to constitutional protections. Those who take this view claim that, at best, email can be given constitutional privacy protections only if it is encrypted. This is only one example of the myriad issues that are emerging as courts have to determine how much of our real world, individual privacy will migrate into cyberspace.

The task the courts face in this regard has only been made more complicated by the events of September 11, 2001. Law enforcement is legitimately concerned with preventing terrorist activity in this country and with prosecuting those who endeavor to inflict terrorism upon us. To that end, law enforcement has sought, and has been given, expanded powers to conduct surveillance, to engage in investigations, to sanction and to punish offenders, and Attorney General Ashcroft has made it clear that these powers will be used to their maximum in an effort to keep the country safe from those who would do it harm. While these powers have been legitimately conferred upon law enforcement, we must maintain vigilance to ensure that they are not abused in a manner that erodes the individual rights to privacy which have for more than a century been a defining characteristic of life in this country. We must also maintain vigilance to ensure that the conferral of these powers does not irresponsibly lead to further expansions in the authority of law enforcement to monitor activity in this country.

## REFERENCES

1. *See, e.g.,* The Bill of Rights, U.S. National Archives & Records Administration, http://www.archives.gove/exhibithall/charters/offreedom/billofrights/amendments1-10.html.

2. *See, e.g., A More Perfect Union: The Creation of the U.S. Constitution,* U.S. National Archives & Records Administration, http://www.archives.gov/exhibithall/chartersoffreedom/constitution/constitutionhistory.html

3. *See, e.g.,* First Amendment: Annotations, FindLaw, http://caselaw.lp.findlaw.com/data/constitution/amendment01/.

4. *See, e.g.,* Fourth Amendment: Annotations, FindLaw, http://caselaw.lp.findlaw.com/data/constitution/amendment04/.

5. *See, e.g.,* Fifth Amendment: Annotations, FindLaw, http://caselaw.lp.findlaw/com/data/consititution/amendment05/.

6. 277 U.S. at 473-474.

7. 389 U.S. at 352.

8. *See* 18 U.S. Code at 2510-2516.

9. *See, e.g.,* Charles Doyle, The Patriot Act: A Legal Analysis, Congressional Research Service CRS-2 to CRS-3(April 15, 2002:, http://www.fas.org/ipr/crs/RL31377.pdf ("When approved by senior Justice Department officials, law enforcement officers may seek a court order authorizing them to secretly capture conversations concerning any of a statutory list of offenses (predicate offenses)"). U.S.C. 2516.9.

10. *See, e.g.,* U.S. Department of Justice, Searching and Seizing Computers and Obtaining Electronic Evidence in Criminal Investigations § III (2002), http://www.cybercrime/gov/s&smanual2002.htm#III.

11. This name is an acronym that summarizes the Act's full title, which is the Uniting and Strengthening America by Providing Appropriate Tools Required to Intercept and Obstruct Terrorism.

12. For a detailed treatment of the Patriot Act, *see* Charles Doyle, *The USA PATRIOT Act: A Legal Analysis,* Congressional Research Service (April 15, 2002), http://www.fas.org/irp/crs/RL31377.pdf.

13. *See, e.g.,* Charles Doyle, The Patriot Act: A Legal Analysis, Congressional Research Service CRS-4 (April 15, 2002), http://www.fas.org/irp/crs/RL31377.pdf. The concept of Title III predicate offenses is discussed in the previous section.

14. *See, e.g.,* Charles Doyle, The Patriot Act: A Legal Analysis, Congressional Research Service CRS-4 (April 15, 2002), http://www.fas.org/irp/crs/RL31377.pdf.

15. U.S. Department of Justice—Computer Crime and Intellectual Property Section, Field Guidance on New Authorities that Relate to Computer Crime and Electronic Evidence Enacted in the USA Patriot Act of 2001, http://www.cybercrime.gov/PatriotAct.htm. See 18 U.S. Code § 1030.

16. *See e.g.,* Electronic Frontier Foundation, EFF, Analysis of the Provisions of the USA Patriot Act (October 31, 2001). http://www.eff.org/Privacy/Surveillance/Terrorismmilitias/20011031effusapatriotanalysis/html.

17. *See, e.g.,* Electronic Frontier Foundation, EFF Analysis of the Provisions of the USA Patriot Act (October 31, 2001). http://www.eff.org/Privacy/Surveillance/Terrorismmilitias/20011031effusapatriotanalysis.html.

18. *See, e.g.,* Electronic Frontier Foundation, EFF Analysis of the Provisions of the USA Patriot Act (October 31, 2001). http://www.eff.org/Privacy/Surveillance/Terrorismmilitias/20011031effusapatriotanalysis.html.

19. *See, e.g.,* Charles Doyle, The Patriot Act: A Legal Analysis, Congressional Research Service CRS-5 (April 15, 2002), http://www.fas.org/irp/crs/RL31377.pdf. *See also* U.S. Department of Justice—Computer Crime and Intellectual Property Section, Field Guidance on New Authorities that Relate to Computer Crime and Electronic Evidence Enacted in the USA Patriot Act of 2001, http://www.cybercrime.gov/PatriotAct.htm (revision of the statute means that "orders for the installation of pen register and trap and trace devices may obtain any non-content information—all "dialing, routing, addressing, and signaling information"—utilized in the processing and transmitting of wire and electronic communications. Such information includes IP addresses and port numbers, as well as the 'To' and 'From' information contained in an e-mail header").

20. *See, e.g.*, U.S. Department of Justice—Computer Crime and Intellectual Property Section, Field Guidance on New Authorities that Relate to Computer Crime and Electronic Evidence Enacted in the USA Patriot Act of 2001, http://cybercrime.gov/PatriotAct.htm ("Pen/trap orders cannot...authorize the interception of the content of a communication, such as words in the 'subject line' or the body of an e-mail").

21. *See, e.g.,* Charles Doyle, The Patriot Act: A Legal Analysis, Congressional Research Service CRS-4 (April 15, 2002), http://www.fas.org/irp/crs/RL31377.pdf.

22. *See, e.g.,* Charles Doyle, The Patriot Act: A Legal Analysis, Congressional Research Service CRS-4 (April 15, 2002), http://www.fas.org/irp/crs/RL31377.pdf.

23. *See* 18 U.S. Code at 3121-3127.

24. Charles Doyle, The Patriot Act: A Legal Analysis, Congressional Research Service CRS-4 (April 15, 2002), http://www.fas.org/irp/crs/RL31377.pdf. *See also* 18 U.S. Code §§ 3121-3127.

25. U.S. Department of Justice—Computer Crime and Intellectual Property Section, Field Guidance on New Authorities that Relate to Computer Crime and Electronic Evidence Enacted in the USA Patriot Act of 2001, http://www.cybercrime.gov/PatriotAct.htm. "Generally, when law enforcement serves a pen/trap order on a communication service provider that provides Internet access or other computing services to the public, the provider itself should be able to collect the needed information and provide it to law enforcement. In certain rare cases, however, the provider may be unable to carry out the court order, necessitating installation of a device (such as Etherpeek or the FBI's DCS1000) to collect the information. *Id.*

26. *See, e.g.,* U.S. Department of Justice—Computer Crime and Intellectual Property Section, Field Guidance on New Authorities that Relate to Computer Crime and Electronic Evidence Enacted in the USA Patriot Act of 2001, http://www.cybercrime.gov/PatriotAct.htm.

27. U.S. Department of Justice—Computer Crime and Intellectual Property Section, Field Guidance on New Authorities that Relate to Computer Crime and Electronic Evidence Enacted in the USA Patriot Act of 2001, http://www.cybercrime.gov/PatriotAct.htm.

28. U.S. Department of Justice—Computer Crime and Intellectual Property Section, Field Guidance on New Authorities that Relate to Computer Crime and Electronic Evidence Enacted in the USA Patriot Act of 2001, http://www.cybercrime.gov/PatriotAct.htm.

29. U.S. Department of Justice—Computer Crime and Intellectual Property Section, Field Guidance on New Authorities that Relate to Computer Crime and Electronic Evidence Enacted in the USA Patriot Act of 2001, http://www.cybercrime.gov/PatriotAct.htm.

30. U.S. Department of Justice—Computer Crime and Intellectual Property Section, Field Guidance on New Authorities that Relate to Computer Crime and Electronic Evidence Enacted in the USA Patriot Act of 2001, http://www.cybercrime.gov/PatriotAct.htm.

31. U.S. Department of Justice—Computer Crime and Intellectual Property Section, Field Guidance on New Authorities that Relate to Computer Crime and Electronic Evidence Enacted in the USA Patriot Act of 2001, http://www.cybercrime.gov/PatriotAct.htm.

32. U.S. Department of Justice—Computer Crime and Intellectual Property Section, Field Guidance on New Authorities that Relate to Computer Crime and Electronic Evidence Enacted in the USA Patriot Act of 2001, http://www.cybercrime.gov/PatriotAct.htm.

33. U.S. Department of Justice—Computer Crime and Intellectual Property Section, Field Guidance on New Authorities that Relate to Computer Crime and Electronic Evidence Enacted in the USA Patriot Act of 2001, http://www.cybercrime.gov/PatriotAct.htm.

34. U.S. Department of Justice—Computer Crime and Intellectual Property Section, Field Guidance on New Authorities that Relate to Computer Crime and Electronic Evidence Enacted in the USA Patriot Act of 2001, http://www.cybercrime.gov/PatriotAct.htm.

35. U.S. Department of Justice—Computer Crime and Intellectual Property Section, Field Guidance on New Authorities that Relate to Computer Crime and Electronic Evidence Enacted in the USA Patriot Act of 2001, http://www.cybercrime.gov/PatriotAct.htm. *See* 18 U.S. Code § 2702(c)(3).

36. U.S. Department of Justice—Computer Crime and Intellectual Property Section, Field Guidance on New Authorities that Relate to Computer Crime and Electronic Evidence Enacted in the USA Patriot Act of 2001, http://www.cybercrime.gov/PatriotAct.htm.

37. *See, e.g.*, U.S. Department of Justice—Computer Crime and Intellectual Property Section, Field Guidance on New Authorities that Relate to Computer Crime and Electronic Evidence Enacted in the USA Patriot Act of 2001, http://www.cybercrime.gov/PatriotAct.htm. See also 18 U.S. Code at 2510 & 2511.

38. *See, e.g.*, Charles Doyle, The Patriot Act: A Legal Analysis, Congressional Research Service (April 15, 2002), http://www.fas.org/irp/crs/RL31377.pdf.

39. Charles Doyle, The Patriot Act: A Legal Analysis, Congressional Research Service (April 15, 2002), http://www.fas.org/irp/crs/RL31377.pdf.

40. *See, e.g.*, Charles Doyle, The Patriot Act: A Legal Analysis, Congressional Research Service (April 15, 2002), http://www.fas.org/irp/crs/RL31377.pdf.

41. Charles Doyle, The Patriot Act: A Legal Analysis, Congressional Research Service (April 15, 2002), http://www.fas.org/irp/crs/RL31377.pdf. *See* 50 U.S. Code 1801et seq.

42. *See, e.g.*, Charles Doyle, The Patriot Act: A Legal Analysis, Congressional Research Service (April 15, 2002), http://www.fas.org/irp/crs/RL31377.pdf.

43. *See, e.g.*, Charles Doyle, The Patriot Act: A Legal Analysis, Congressional Research Service (April 15, 2002), http://www.fas.org/irp/crs/RL31377.pdf

44. Nancy Chang, The USA Patriot Act: What's So Patriotic About Trampling on the Bill of Rights?, Center for Constitutional Rights (November 2001), http://www.ccr-ny.org/whatsnew/usapatriotact.asp.

45. *See, e.g.*, Susan W. Brenner, The Voice of the Community: A Case for Grand Jury Independence, 3 Virginia Journal of Social Policy & The Law 67 (1995), http://www.udayton.edu/~grandjur/recent/lawrev.htm. *See also* Grand Jury Secrecy, Federal Grand Jury Website, http://www.udayton.edu/~grandjur/feedback/nav/secrecygj.htm.

46. *See, e.g.*, Susan W. Brenner, The Voice of the Community: A Case for Grand Jury Independence, 3 Virginia Journal of Social Policy & The Law 67 (1995), http://www.udayton.edu/~grandjur/recent/lawrev.htm. *See also* Rule 6 of the Federal Rules of Criminal Procedure, http://www.law.ukans.edu/research/frcrilll.htm; Grand Jury Secrecy, Federal Grand Jury Website, http://www.udayton.edu/~grandjur/feedback/nav/secrecygj.htm.

47. *See, e.g.*, Susan W. Brenner, The Voice of the Community: A Case for Grand Jury Independence, 3 Virginia Journal of Social Policy & The Law 67 (1995), http://www.udayton.edu/~grandjur/recent/lawrev.htm.

48. *See, e.g.*, Susan W. Brenner, The Voice of the Community: A Case for Grand Jury Independence, 3 Virginia Journal of Social Policy & The Law 67 (1995), http://www.udayton.edu/~grandjur/recent/lawrev.htm. *See also* Grand Jury Secrecy, Federal Grand Jury Website, http://www.udayton.edu/~granjur/feedback/nav/secrecy.gj.htm.

49. *See, e.g.*, Charles Doyle, The Patriot Act: A Legal Analysis, Congressional Research Service CRS-21 (April 15, 2002), http://www.fas.org/irp/crs/RL31377.pdf.

50. *See, e.g.*, Charles Doyle, The Patriot Act: A Legal Analysis, Congressional Research Service CRS-20 (April 15, 2002), http://www.fas.org/irp/crs/RL31377.pdf.

51. Charles Doyle, The Patriot Act: A Legal Analysis, Congressional Research Service CRS-21 (April 15, 2002), http://www.fas.org/irp/crs/RL31377.pdf (quoting Rule 6(3)(C)(i)(V) of the Federal Rules of Criminal Procedure).

52. Charles Doyle, The Patriot Act: A Legal Analysis, Congressional Research Service CRS-21 (April 15, 2002), http://www.fas.org/irp/crs/RL31377.pdf.

53. *See e.g.,* Fourth Amendment: Annotations, FindLaw, http://caselaw.lp.findlaw.com/data/constitution/amendment04/. *See also* Rule 41 of the Federal Rules of Criminal Procedure, http://www.law.ukans.edu/research/frcrimIX.htm.

54. *See, e.g.,* Charles Doyle, The Patriot Act: A Legal Analysis, Congressional Research Service CRS-63 (April 15, 2002), http://www.fas.org/irp/crs/RL31377.htm.

55. Charles Doyle, The Patriot Act: A Legal Analysis, Congressional Research Service CRS-63 (April 15, 2002), http://www.fas.org/irp/crs/RL31377.htm.

56. *See, e.g.,* George A. Chidi, Jr. FBI Claims Keystroke Logger Is National Secret, Network World Fusion (August 27, 2001), http://www.nwfusion.com/news/2001/0827fbikey.html; Robert Vamosi, We Know What You're Typing…and So Does the FBI, MSNBC News (December 7, 2001), http://www.msnbc.com/news/669010.asp.

57. Donald E. Wilkes, Jr., Sneak and Peek Search Warrants, Flagpole Magazine (September 11, 2002), http://www.law.uga.edu/academics/profiles/dwilkesmore/36sneak/html.

58. Donald E. Wilkes, Jr., Sneak and Peek Search Warrants, Flagpole Magazine (September 11, 2002), http://www.law.uga.edu/academics/profiles/dwilkesmore/36sneak/html.

59. Donald E. Wilkes, Jr., Sneak and Peek Search Warrants, Flagpole Magazine (September 11, 2002), http://www.law.uga.edu/academics/profiles/dwilkesmore/36sneak/html.

60. *See* 18 U.S. Code § 3103a (as amended by the Patriot Act).

61. *See* 18 U.S. Code § 3103a (as amended by the Patriot Act).

62. *See* 18 U.S. Code § 3103a (as amended by the Patriot Act).

63. *See, e.g.,* Nat Hentoff, Burglars with Badges: The Return of "Black Bag Jobs," Village Voice (December 3, 2001), http://villagevoice.com/issues/0149/hentoff.php.

## About the Author:

Susan W. Brenner, NCR Distinguished Professor of Law and Technology at the University of Dayton School of Law, is an expert in cybercrime. She writes, speaks and consults in this area. Her internationally known website, http://www.cybercrimes.net, provides information on cybercrime and links to some of her articles.

Chapter 6

# SOME SECURITY ISSUES AND CONCERNS IN THE FINANCIAL SERVICES INDUSTRY

Dan Schutzer
*VP and Director of External Standards & Speech Technology, Emerging Technologies Group, Corporate Technology Office, Citigroup*

**Abstract:** Because of the high value and volume of transaction in financial services, the industry is totally dependent on computers and communications and particularly concerned with cyber security. This paper provides an overview of the major security issues in financial services together with a survey of the measures the industry is taking to provide convenience to its customers while preserving acceptable levels of risk.

**Key words:** Security issues, Financial Services Industry

## 1. INTRODUCTION

The main security concerns of the financial services industry are threefold:

1. Prevention of service denial
2. Protection of sensitive information
3. Protection against fraud and identity theft.

We are concerned with protecting our systems from disruption and service denial attacks. This includes the ability to challenge and authenticate potential users of system resources and services, and the ability to keep records of requests and responses for services that can be used in non-repudiation cases.

We are concerned about protecting our customer's sensitive data, such as their bank account balances, account numbers, passwords, from unauthorized parties, and protecting sensitive company information, such as trade secrets from prying eyes and ears.

Finally, we need to protect our customers from fraud, loss of privacy and identity theft. That is, we must prevent fraud and identity theft from occurring, and if they do occur we must prevent our customers from suffering financial loss, embarrassment, and other inconveniences. This means we have to be willing to absorb such losses, and to assist our customers in repairing their reputation and credit rating if their identity has been stolen and misused.

## 2. SECURITY TOOLS

This chapter provides a survey of approaches used by financial service firms in their never ending struggle to cope with the three sets of security concerns. To achieve these goals we employ a variety of tools ranging from firewalls to identity management. Some of the key tools are discussed below:

- We use *firewalls* to prevent unauthorized access and to filter out unwanted services and sites.

- *Anti-virus software* is used to detect and contain and purge the system of viruses.

- *Anti-spam software* is used to filter out unwanted email solicitations that could otherwise clog up the system, direct the user to undesirable sites, and carry virus attachments.

- *Anti-site spoofing* techniques are used to allow a user to be sure they are at a desired site and not a false site spoofing the real site that is attempting to lure the user into providing private or sensitive information to the wrong site.

- We employ software to *monitor and spot suspicious activity.* This software looks for anomalous behavior activity, attempts to break-in to the system, and other such signs of suspect activity, and provides *warnings and alerts* in an effort to prevent or detect as early as possibly fraudulent or criminal behavior, tampering, service denial attempts or system attacks. Additional customer confirmation and verification mechanisms are used when suspicious and anomalous activity is detected to ensure only authenticated and authorized users can access the system.

- We employ tools and processes for reconstituting service, resolving disputes, catching and prosecuting criminals, preventing further fraud, tampering and service denial attempts, and restoring customer losses. *Alternate remote back-up* sites are established, and *contingency plans* are developed to maintain Continuity of Business (COB) under calamities that range from natural disaster to terrorist attacks. We employ people, processes and technology to aid us in *tracing, tracking and active entrapment of criminals*

- *Audit trails* are maintained to keep records that can be used to reconstruct transactions when the system fails, to resolve disputes, and to enforce non-repudiation.

- *Cryptography* is used for a number of purposes. Encryption is used to maintain confidentiality and to protect sensitive information, such as passwords [3]. Public key cryptography is used for electronic distribution of the encryption key, and for digital signatures to prevent tampering and for non-repudiation [3].

# 3. MAINTAINING A SECURE ENVIRONMENT

There are many issues associated with maintaining a secure, robust environment. Some of the main issues are discussed below:

- Patch maintenance
- Insider threat
- Increased dependence on outsourcing and generic vendor packages
- Lack of control over transacting device
- Backdoors and social engineering
- Identity Management and Authentication.

One issue involves the cost and risks associated with keeping up with all the vendor patches and updates. When vulnerabilities are discovered, it is important to develop, distribute and install systems patches that fix these vulnerabilities as soon as possible. However, although vendor patches and updates are designed to fix reported bugs and vulnerabilities, they can often introduce new instabilities and vulnerabilities, so they must be thoroughly tested and assessed. The frequency at which vulnerabilities are discovered and patches developed is increasing at such a rate, that if we tried to keep our systems up-to-date with patches as they occur, we would find ourselves in a constant state of change and update. Yet we still have trouble countering new threats and new methods of exploitation. We need systems that can dynamically detect, isolate and adapt to new threats, much like the way the human immune system works.

Another issue involves the insider threat, the vulnerability of our systems to insiders. This vulnerability must be managed through careful employee background checks and monitoring, institution of a system of checks and balances, strong identity management, and detection of unusual patterns of use.

Financial firms, and our customers, increasingly outsource large portions of their operations and rely on packaged software solutions, in the interest of cost and efficiency. This results in increased dependence of our systems and operations on the security and robustness of third parties outside our control. It necessitates our careful auditing of these third parties and inclusion in our contracts with the third parties of the necessary safeguards. It has also opened up a critical review of the liability and responsibility of vendors for their products' security flaws. As an illustration, a few days ago, somewhere around 60% of the major backbone routers on the Internet were taken out of service for about 1 ½ hours due to a viral attack. Although this had nothing to do with the bank's systems, we were impacted. To a certain extent, we succeeded in this instance to minimize the impact by investing, at considerable cost, in alternate back-up sites and communications.

Our customers increasingly transact using commodity devices over which we have no control with regard to making them more secure and less vulnerable to attack. For example, customers increasingly use their own PC's, phones and Palm devices to communicate with us over networks provided and operated by vendors and third parties. We could run our own networks and provide our customers with more secure phones and PDAs of our own design, but this would not be practical when we are talking about the large numbers of customers we have. For example, with 80 million customers, absorbing even a cost of $10 per customer is not very cost-effective. And the transacting devices become less valuable to our customers if they only work over our networks and are only able to access our services. There is no general solution to this problem, but the financial services community is exploring two approaches in parallel; namely:

1. Publishing security standards and testing certification, in concert with partners like the U.S. Government and ANSI and other standards bodies, to help put some pressure on vendors.

2. Exploring ways in which we can cost-effectively add the necessary security overlaid on top of existing devices.

Since criminals often find the quickest easiest path, which usually involves "social engineering", namely tricking our customers and employees into giving them valuable information, we must stay alert and keep our employees and customers educated on the kinds of tricks to expect, and what not to do. To this end, most financial service firms have established a Corporate Security Office, with satellite offices in each business. The role of the Corporate Security Office (CSO) is to serve as an independent agency over the Corporation to ensure that we maintain the highest levels of security. The CSO monitors all vulnerabilities and oversees both technology and policy and procedure fixes to maintain security amongst our systems and services. Among other things the CSO's role is to collect and monitor these social engineering activities, and publish corporate policy and guidelines to the businesses, who in turn educate their employees and customers.

Identity Management involves, managing our customers' and employees' authentication credentials (e.g. passwords) and entitlements, permissions and authorizations to our various web and legacy applications. This includes keeping track of the systems privileges granted to a customer or employee, making the necessary changes as an employee's position and responsibility changes, or when the customer leaves, or an employee is terminated. This capability must also include tools to assist in updating systems provisioning, that is updating all the various systems to reflect changes in user profiles, authentication credentials, and entitlements/permitted actions. This can be especially costly for our employees who typically have access to many different systems. The number of systems that a typical employee needs to sign onto can be as high as 10 or more (e.g. email, voice mail, Intranet, HR Personnel file, 401 K pension, health plan, procurement system, various work applications). Managing, synchronizing and keeping these multiple systems up-to-date can prove time consuming and costly. Today this process is heavily manpower intensive. Most financial service firms are implementing Single Sign-on solutions, discussed below, to help reduce this overhead and make the systems more manageable without sacrificing security.

# 4. SINGLE SIGN-ON SOLUTIONS

Recently, there has been a strong movement to introduce Single Sign-On (SSO) solutions for both our customers and employees [5]. This is in reaction to the increasing number of authentication credentials (today, this typically means a large number of passwords, 4 or more), that an individual needs, and the overhead involved in maintaining all these passwords. Allowing for a single authentication credential that provides an employee access to all their systems and authorizations; or that provides a customer access to all our products and services makes it more convenient for the individual to access their accounts/services and reduces the incidence of someone forgetting their password and having to endure the cost and overhead associated with processes such as password reset. However, it makes our customers' and employees' authentication credentials more valuable to a fraudster and a more attractive target, as it holds the keys to all of the customer's accounts, permissions and entitlements.

Because a single authentication credential can now allow the user access to a greater number of systems and entitlements, there is interest in moving to stronger authentication technologies (such as tokens, smart cards and biometrics [4].) Often these techniques are combined to provide multiple factor authentication (e.g. more than one authentication credential, such as possession of a token, something one has, a biometric, such as a fingerprint, an attribute that is associated with the individual, and a password, something that one knows), and tiered authentication—where the number and forms of authentication required varies with the risk profile of the associated business activity.

The need for stronger authentication motivates us to incorporate other supporting authentication technologies, such as anomaly detection and the use of collateral information such as cross-referencing and checking customer out-of-wallet information. Is the system being accessed at a time and location that is consistent with past behaviour? Is the customer dialling in from a known number? Do the inquiries and actions fit with past behaviour? If one asks enough different personal questions, it becomes increasingly unlikely that an impersonator will know the correct answers to all the questions.

Support for tiered authentication, authorization and access controls is needed so we can match the authentication strength with the risk. Security generally involves a trade-off between increased cost and inconvenience and increased security. Since security controls generally grow in cost and inconvenience as they grow in strength, we generally require the strongest security and authentication technologies only when the risk warrants it.

The Single Sign-On capability is also being extended cross enterprises; e.g. from corporation-to-corporation. So, for example, an employee at General Motors can log onto their General Motors Intranet, and from their Intranet, can reach the Citigroup 401K pension plan site for General Motors employees, without needing to log-on and authenticate themselves to the Citigroup site. Instead, Citigroup relies upon General Motors authenticating their own employees and vouching for these employees to Citigroup by means of the exchange of identity credentials. The same can be done for procurement services, corporate cards and so on. This same capability is being contemplated in the consumer world, where a consumer can switch from various partner sites (e.g. their credit card company, car rental company and airline sites) without the need for re-authentication. This represents, not just a technical challenge, but requires a large degree of trust between corporations and some well-thought out liability sharing amongst the corporations, so that SSO functionality can work in a smooth, safe, secure manner. With respect to liability sharing, it's a two-way street. Is a corporation willing to authenticate one of its customers for a third party, and vouch for that authentication? Alternatively is the corporation willing to accept the authentication of one of their customers by another third party? At the moment, there is no universal rule for when to take the risk - it's a very gray area. The decision as to when to accept this risk depends upon a number of things; namely: the potential revenue; the business relationship; the degree of trust between the two corporations; the clout of each of the two corporations, and how well each corporation feels it can manage the risk.

## 5. TRADE-OFF

For a financial service firm, ensuring a secure, robust environment always involves a trade-off; that is generally, the more secure and robust the system, the greater the cost and the greater the inconvenience to the customer and employee with respect to accessibility and ease-of-use. Generally, the more secure we make something, the greater the cost and time required, and the more limits and constraints placed on the user.

Security concerns tend to inhibit the introduction of new functions and services. The more frequently we introduce new functionality and features, the more often we can inadvertently introduce new vulnerabilities if we are not careful to do thorough enough testing. The more we make our customers' lives simpler through the use of technology, the more dependent we make our customers on these technology solutions being available and running 24x7. It is very important to get this trade-off right, in the sense that if a system is too inconvenient, and/or too costly, no one will use it; whereas if it is too insecure, and too readily accessible without necessary checks and balances, then the risk exposure grows, and the potential for damage to our brand increases as the chances for fraud, service denial, and bad press increase.

A business always has to make a trade-off between increasing the cost and complexity of their services and reducing the risk by increasing the security. Business will generally weigh this trade-off very carefully, and oftentimes they will opt for doing things that will keep costs down, and increase the convenience to the customer first, even if the risk increases and the security decreases. They will try to compensate for this increased risk, through processes and procedures, and through greater vigilance. Likewise, businesses are more likely to choose to invest in additional functionality that leads to more revenue and attracts new customers, long before they invest in more security, as long as they feel they can manage and contain the risk. Generally, financial services organizations will only invest in security if the cost of the investment is much less than the value of the resultant decrease in risk exposure and potential losses.

Because a corporation's financial system is usually dependent upon the systems of other corporations and associations, even if we invested heavily in securing our own systems and processes, we are often no more secure than the systems we are interconnected to, and dependent upon. For example, if one bank fails during the day in the financial clearing and settlement system (e.g. the bank can't make one of their payments or transfer stock), this will impact the entire network and all the other corporations. Likewise, if the communications networks we employ fail to maintain a quality connection between our systems and our customers' systems, then our systems fail. So, to adequately secure our systems requires a great deal of cooperation amongst the various players in the financial community and their service providers. There are several financial industry and cross-industry initiatives in this area that are attempting to first identify these vulnerabilities and then to cooperate in solutions to fixing them.

# 6. DISTRIBUTION CHANNELS

In our eagerness to offer our customers increased convenience and to allow them to transact with us anywhere, anyway, anytime, we are constantly exploring new distribution channels over which to deliver our products and services as well as new functionality. This includes such innovations as delivery of our services and products over the Internet; over cellular phones and other wireless, mobile communications; over interactive TV; and via web services.

Each innovation raises its own set of security concerns, challenges, and advantages. For example, the idea of getting real-time notifications (such as notification that a bill is due, that a balance or credit limit has crossed some threshold, or that there is suspected fraud), and being able to take action on these notifications over something like a personal cell phone or wireless palm device is quite appealing. It has the advantage that we can be more confident that a message sent to one's cellular phone is more likely to reach them. One is not always in one's office or home, and someone else can be in the home and available to answer an incoming call or read an incoming email, if the desk-top computer is left on. It is more likely that the cell phone, or PDA, can be linked to an individual, whereas the home phone, or PC, is generally a shared family appliance. However, there are many security concerns with mobile phones and PDA's. Communications are more easily intercepted and often use less secure encryption that gets broken out in the clear at more points along the transmission path, and the hand-held devices often store sensitive information without adequate protection. For example, many sensitive Short Message Service (SMS) messages and emails are stored in the clear on the phone, and can be accessed without the need for a password. So, if the phone falls into the wrong hands, lots of sensitive information becomes available to the wrong person.

The Internet raised the security bar, increasing vulnerability to hackers, susceptibility to site spoofing, greater exposure to customer transaction devices not under our direct control, and greater dependence on third party web sites (e.g. web merchants) who store sensitive customer information, such as our customers' credit card numbers. The Internet has necessitated the need for ethical hacks, and other innovations, such as one-time-alternate credit card numbers and digital wallets to protect the customer's card and account numbers, digital certificates and other means of allowing our customers to be sure they are talking to us rather than some imposter when they access us over the web. As more of our infrastructure adopts web technology, such as our telephone networks embracing Voice Over IP

(VoIP) [7]) then we must remain cautious as to any new security risks we are introducing.

Recent terrorist threats and increasing cyber-terrorism has placed added emphasis, and increased attention and guidance from regulators. For example, the Office of the Comptroller of the Currency (OCC) [6] and Securities Exchange Commission (SEC) require critical infrastructures, such as the Financial Service Infrastructure, to have sufficient back-up, including alternate centers and communications paths and continuity of business contingency plans. Initial emphasis is being placed on the large wholesale clearing and settlement systems, but ultimately, one would expect attention might move on to the consumer front (e.g., ATM and credit card networks).

This includes the need for Financial Services firms to respond to new governmental regulatory and reporting needs involving monitoring, detecting and reporting on suspicious possible terrorist and money laundering activities, and on the integrity and security of the financial systems. Many of these reporting requirements vary state-by-state and country-by-country. This makes life difficult and more costly for a large global company.

These reporting requirements often are in conflict with our privacy obligations to our customers, and even related governmental privacy laws and guidelines. It raises issues regarding both the need to share information with other companies from the same industry, with related industries we are dependent upon, and with the governments. But this sharing of information can, under the wrong circumstances, cause increased liability and exposure. Will this sharing make vulnerabilities more widely known and more easily exploitable? Will they result in bad press and loss of customers? Can we be certain that privileged information will not be revealed to the public either inadvertently, or as part of some conflicting law, such as the Freedom of Information Act (FOIA) [1]. As an example of new guidance and reporting requirements with financial impact, financial services firms are being asked to quantify their operational risk, and even to set aside financial reserves to cover future potential losses, that are based upon this assessment of the operational risk.

## 7. NEW TECHNOLOGIES

Technology keeps advancing, and in an attempt to allow our customers to do business with us as conveniently as possible, any way, any time, financial service firms are always exploring the development of new channels for their customers to interact and transact with them. Every time we embrace a new technology to create a new communications channel for our customers it opens up a new set of security concerns. By its very nature, a new technology introduces a new set of unknowns. We have no experience with using the technology and haven't yet discovered all its vulnerabilities and attack points. Moreover when a technology is immature, its first implementation often does not have security concerns first and foremost in the minds of its developers. They are more interested in creating a new marketplace by delivering functionality for an affordable price, not whether the implementation is secured against all possible forms of attack. These points can be best illustrated by examining some examples of relatively new technologies that have been piloted for use in delivering financial services to our customers.

Cellular phone service has taken the world by storm. It has already been hugely successful as a means for customers to communicate with each other while on the move. Primarily a voice channel, the cellular phone industry has also introduced data communications services over the cellular phone network, making it appealing for financial service firms to extend their services over the cellular phone network, where millions of their customers can already be found. However, delivery of services over cellular phones adds a new set of security concerns. One concern deals with the way security and encryption is currently handled including the introduction of 2-zone security models. The lower channel bandwidth and device interface limitations (e.g. limited compute power and end-user display and input device capability) have generally led to less capable and secure encryption and authentication technologies and methodologies, making services less secure when delivered over these channels. Adding to this concern, because the security is not as strong as over land lines, when the cellular data is encrypted (it often is not encrypted at all), it is generally decrypted and then re-encrypted by the cellular carrier at the point it connects into the landline network. This leaves sensitive information flowing between the customer and the financial service firm, in the clear at some third party premises. It has created the need to review the security of the carrier's systems, processes and premises and caused us to develop special service level agreement (SLA) terms including terms that ensure adequate protection of information at the carrier, acceptable encryption, and the existence of a persistent, reliable identifier (e.g. MSISDN or equivalent [2].) There is also a need to assess and evaluate the telecommunication gateways, as well as to establish gateways within our own premises that ensure that adequate security provisions are met and that we can simplify the complicated navigation and configuration options and non-interoperable systems in the cellular world.

Additionally, the mobile phone often itself can introduce security concerns, requiring us, for example, to address the need to prohibit echoing and local storage of sensitive information in the clear on the device. Another concern is using the Short Message Service (SMS) messaging system, a very popular service, to communicate with our customers. SMS transactions are stored in the clear on the cell phone, so if the phone is lost and stolen, or if this information can be echoed or inadvertently sent to some third party, then any sensitive information in an SMS message can be easily discovered.

Fixed wireless, 802.11 type channels make it easy for interlopers to remotely and wirelessly piggyback on our networks. This introduces the need to counter potential IP spoofing, so that we know who is on our networks and what data exchanges they are permitted to see.

Cable modems introduce new vulnerabilities, regarding the storage of sensitive information at the cable head-end, and the need to encrypt data over a cable that is potentially shared/seen at all the houses connected by the same cable.

Voice over IP (VoIP) and IP Telephony introduces a new set of challenges as well. Suddenly our phone networks, become vulnerable to a new set of threats. Packet flooding can create service denial. For example, without statefull firewalls, a User Datagram Protocol (UDP) flood denial of service attack launched from the data segment, could easily overwhelm the voice segment.

As IP telephones, are nothing more than PC's they are especially susceptible to attack. They can be turned into rogue devices that can pose serious threats to the network. Telephony devices generally don't support confidentiality and encryption.

Firewalls can be defeated when VoIP protocols require a hole in the firewall for streaming audio to pass through.

To defeat many of these attacks, establishing identity of the device and the source of the packet, etc. is key. In other words, if we knew that data was being received from a known and authenticated trustworthy device, we could safely open up our firewalls to that device for many of these additional services.

# 8. IMPACT OF TERRORISM

Finally, new threats (e.g. Physical Terrorism attacks coordinated with Cyber Terrorist attacks) introduce the need for new innovative approaches to counter these threats. We are exploring a number of approaches. Some illustrative examples are discussed below.

One area of investigation is whether we can develop and invest in rapidly constituted communications assets. For example, portable radio transmitters that can be quickly and easily set up outside buildings or on balloons that can be launched quickly and inexpensively.

We are establishing alternative processing centers far away from urban areas, with diverse communications into and out of the area. These alternate processing centers, also need contingency plans developed for the rapid deployment of critical personnel to these centers.

Prioritization and reservation of surviving communications in an emergency is another way of staying connected to conduct absolute minimum essential communications, and the need to keep financial service firms connected is an essential activity in times of crisis. The establishment of processes and contingency plans to enable operation with minimum essential communications and processing capacity is also a requirement.

Also being explored is the concept of a safe mode operation that a system can fall back on in case of service disruption, and the introduction of adaptive self-healing systems techniques to keep the remaining assets in our distributed systems after an attack running, even if at reduced capacity.

We also are studying the use of biometrics as a means of strengthening our identity authentication. But, this raises many new issues as the various biometrics have a whole new class of performance issues (e.g. the possibility of false reject and false accept), and a whole new set of possible attacks. The worst thing we can do is to think we are more secure while in actuality introducing a whole new set of vulnerabilities.

# 9. CONCLUSION

The financial services firm takes security very seriously. We recognize that the threat is constantly changing, and like an opponent in a chess match, we have to be constantly vigilant of new threats, new attack approaches, and new, previously undiscovered, vulnerabilities in our systems that need to be fixed.

An added complication is the ever changing, constant need to improve and add to our service offerings and supporting infrastructure, in an effort to stay competitive in our industry and to serve our customer better, faster, and more inexpensively. This means that many of our existing security plans and approaches have to be constantly re-examined and tested, not just because new threats emerge, but because we ourselves are constantly introducing new systems and new procedures, and thus risking the introduction of new unknown vulnerabilities, in an effort to stay competitive. There is no magic bullet, just the need to constantly monitor, prevent existing known attacks, and detect, analyze, isolate and fix new threats and vulnerabilities as they arise.

## REFERENCES

1. Freedom of Information Act. http://foia.state.gov/
2. Transmission of the MSISDN number to the application. zhttp://www.aeif.org/db/docs/ccm/Annex_to_037-0022a.pdf
3. Cryptography. http://www.cryptography.com/
4. Authentication Technologies. http://www.rsasecurity.com/products/securid/
5. Single Sign-On Solutions in a Mixed Computing Environment (Timo Tervo). http://www.tml.hut.fi/Opinnot/Tik-110.501/1998/papers/5singlesign-on/singlesignon.htm
6. Office of the Comptroller of the Currency (OCC). http://www.occ.treas.gov/
7. Voice over IP (VOIP). http://www.cis.ohio-state.edu/~jain/refs/ref_voip.htm

**About the Author:**

Daniel Schutzer is the Vice President & Director of External Standards and Speech Technology, Emerging Technology Group, Corporate Technology Office, CitiGroup. He is also Financial Service Technology Consortium, Board Chairman, Chairman of ISO Subcommittee 2, Fellow and Advisory Board member of the New York Academy of Sciences, on the BITS Advisory Group, an IFX Board Member, OFX Banking Co-Chairman, and Citibank ANSI X9 representative. Current areas of focus include electronic payments and electronic commerce, speech processing (recognition and verification), authentication, fraud detection and control, and security over computer networks.

Chapter 7

# SECURITY AND INSECURITY OF WIRELESS NETWORKS

R. Chandramouli
*Department of Electrical and Computer Engineering*
*Stevens Institute of Technology*

**Abstract:** This chapter discusses some popular approaches in providing wireless network security. Several security features supported by popular wireless standards are presented. Performance factors that are traded off by these standards for creating and maintaining secure communications are explained. Finally, some known and potential security vulnerabilities in current wireless standards are presented along with some challenges and open issues concerning future wireless networks.

**Key words:** Wireless networking, security, IEEE 802.11, encryption

## 1. INTRODUCTION

Wireless communication devices and networks have seen a rapid rise in recent years. The impact of different technologies can be felt in wireless wide area networks (WWAN), local area networks (WLANs) and personal area networking (WPAN). Figure 1 shows the categorization of present day wireless technologies. Clearly, we see from this figure that, different wireless technologies cater to different applications depending on physical and other constraints. For wireless WWANs the coverage area is in kilometers or at least hundreds of meters. High power transmission is required to communicate over such a long distance and the frequency band is licensed. A tradeoff between the high power consumption and data rates is a major concern in WWANs. For WLANs, coverage could vary from 10m to a few hundred meters. Thus, the transmission power levels are lower than that of WWANs because of the smaller coverage distances. Also current WLANs operate in unlicensed frequency bands. Because LANs often are used for relatively high-capacity data communications, they often have fairly high data rates. For instance, IEEE 802.11b WLAN has a nominal range of 100 meters and data rates up to 11 Mbps. This combination of coverage and data rate leads to moderate to high power consumption. Finally, WPANs cover distances in the order of 10 meters or less. This peer-to-peer communication does not usually require exceedingly fast data rates. Bluetooth wireless technology, for example, has a nominal range of 10 meters with raw data rates up to 1 Mbps. The short range and relatively low data rates result in low power consumption, making WPANs technologies suitable for use with small, mobile, battery-powered devices such as PDAs, mobile phones, and digital cameras. In addition, similar to the WLANs, low-power transmission allows for the use of unlicensed frequency bands.

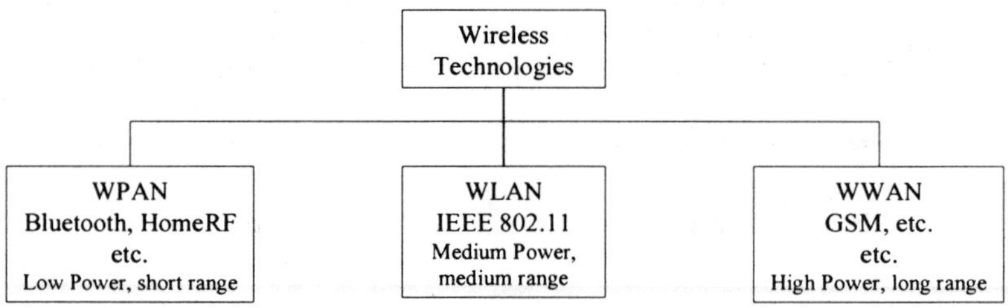


Figure 1. Categorization of wireless technologies

Increasingly, users have started to depend on (mobile) wireless networking for applications such as e-commerce, remote logins, etc., that inherently require privacy and security (e.g., data integrity and authentication) features. Security has become a major concern for these applications because original designs of wireless network architectures did not have security as a top priority. Also, ongoing efforts for the seamless integration of wireless and wired networks produce additional security concerns.

There are similarities and also major differences between wireless and wired network security. Some of the major factors that distinguish wireless security from its wired counterpart are the following:

- Mobility in wireless communications calls for location-dependent security protocols.

• Small wireless devices such as personal digital assistants (PDAs) are constrained by low processing power and memory capacities. This makes it difficult to implement complex, more secure security algorithms.
• Wireless channels are highly unreliable with randomly varying network capacities. Therefore, adaptive security protocols may be necessary.

One of the main security issues in wireless networks occurs when a wireless node radiates energy during transmission that exceeds the allowable physical boundaries. When this happens, an eavesdropper perhaps sitting in the building's parking lot where the wireless node is located, can retrieve the transmitted data using the same wireless network interface card as the transmitting node. Of course, this is possible in a wired LAN also, but only to a lesser extent. In a wired LAN, an eavesdropper would have to be much closer to the cable to intercept the electromagnetic waves radiated through them. We see that there are some common issues in wired and wireless network security, such as:

• Physical layer attacks such as jamming a wireless signal or removing a transmission cable leading to denial of service attacks
• Unauthorized network access and interception
• Attacks from an "insider," i.e., an authorized user attacking the network from within the system

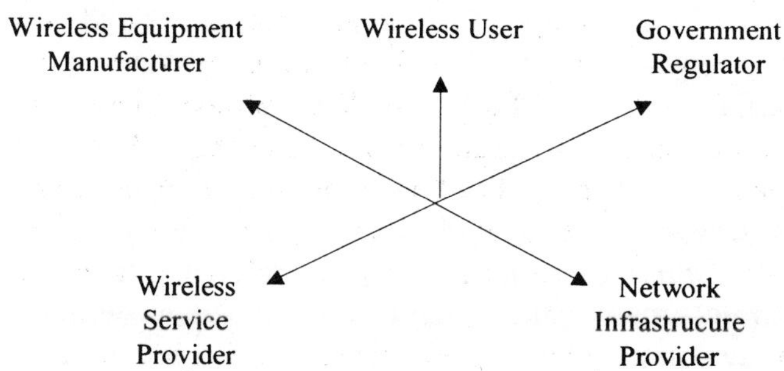


Figure 2. Wireless system security entities

Current wireless system security solutions are not quite effective. One reason that could be attributed to this is the lack of cooperation among the various entities shown in Figure 2. This is because of competing technologies and markets, lack of understanding by a common user of what real security means, not enough business incentives to improve security, and

the lack of a good revenue sharing strategy among network infrastructure providers, wireless service providers, and equipment manufacturers.
In this chapter, we discuss some fundamental issues and challenges in wireless (in)security. An overview of the current approaches to wireless security is given in Section 2. Section 3 describes some of the major security loop holes in wireless technologies and countermeasures. Key open issues are discussed in Section 4.

# 2. OVERVIEW OF APPROACHES TO WIRELESS SECURITY

There are several approaches to wireless security. In this section we provide a brief overview of some the popular ones.

## 2.1 Wired Equivalent Privacy (WEP) Protocol

Since wireless communications use airlink as a shared medium, the transmitted data can be easily intercepted, or unauthorized data can be injected into the network. Therefore, standards such as the IEEE 802.11 recommend link layer security protocols. Link layer security protocols attempt to prevent link layer eavesdropping. The WEP protocol aims at making the wireless security essentially equivalent to a wired access point by encrypting the network traffic and authenticating the wireless nodes. That is, transmitted data are encrypted using the WEP protocol in order to secure the wireless link between a client and an access point, and each authorized user also receives a key thereby disallowing an unauthorized user without a valid key from accessing the network. For a system using this encryption plus authentication combination for security, the following three factors play a major role [12]:

- **User's need for privacy:** How strong the protocols need to be and its cost (MIPS, dollars, and time)
- **Ease of use:** A very difficult-to-use security implementation will not be used.
- **Government regulations:** Encryption is viewed as munitions by many governments, so all encryption products are export controlled.

The WEP protocol tries to balance these general design criteria. WEP encryption begins with a secret key that has been distributed to cooperating stations by an external key management service. WEP encryption is symmetric in that the same key is used for both encryption and decryption. The secret key is combined with an initialization vector (IV) and the resulting seed is input to a pseudo random number generator (PRNG). The PRNG outputs a key sequence of pseudo-random bits equal in length to the largest possible MAC service data unit (MSDU). Two processes are applied to the plaintext MSDU. To protect against unauthorized data modification, an integrity algorithm operates on the plaintext to produce an integrity check value (ICV). Encipherment is then accomplished by mathematically combining the key sequence with the plaintext. The output of the process is a message containing the resulting ciphertext, the IV, and the ICV.

The ultimate selection of the WEP protocol was based on the following specific criteria that were to be met for wireless communications:

- **Reasonably strong:** The security afforded by the algorithm relies on the difficulty of discovering the secret key through a brute-force attack. This in turn is related to the length of the secret key (usually expressed in bits) and the frequency of changing keys. However, it may be an easier problem to discover the key through statistical methods if the key sequence remains fixed and significant quantities of ciphertext are available to the attacker. WEP avoids this by frequently changing the initialization vector and hence the secret key sequence.
- **Self-synchronizing:** Provided by an initialization vector (IV); this property is critical for a data-link level encryption algorithm, where "best effort" delivery is assumed and packet loss rates can be high. An algorithm that assumes reliable delivery in order to maintain synchronization between sender and receiver would not provide acceptable performance.
- **Computationally efficient:** The WEP algorithm is very efficient in comparison to traditional block ciphers. It uses fewer resources and can be implemented efficiently in either hardware or software.
- **Exportability:** As discussed before, government regulations on cryptographic products can be problematic. WEP was designed to be exported to other countries also.
- **Requirement for 802.11 is optional:** Because 802.11 products can be marketed internationally, usage of the WEP algorithm is specifically proposed to be an optional portion of the 802.11 standard.

## 2.2 Wireless Application Protocol (WAP)

Wireless application protocol [10] is an application environment and a set of communication protocols for wireless devices designed to enable manufacturer, vendor, and technology-independent access to the Internet and other advanced services. Originally founded by the WAP forum in 1997 consisting of Ericsson, Nokia, Motorola, and Unwired Planet, WAP is now a global standard. It allows WAP-compliant nodes to interact with other devices and resources. WAP also bridges the gap between the mobile network and the Internet as well as intranets by offering the ability to deliver an unlimited range of mobile services to users independent of their network. Its initial purpose was to define an industry-wide specification for developing applications over wireless communication networks.

The WAP specifications define a set of protocols in application, session, transaction, security, and transport layers, which enable operators, manufacturers, and applications providers to meet the challenges in advanced wireless service differentiation and services. WAP also specifies the wireless transport layer security (WTLS) protocol that is similar to the transport layer security protocol in the Internet. WTLS is responsible for providing authentication, data integrity, and privacy for WAP applications. In WAP, both the client and the server are authenticated, and the connection is encrypted. Man-in-the-middle attacks that can modify data during the transfer is prevented. Although the traffic in the air is encrypted in several mobile networks, the complete end-to-end security is not provided by the mobile network. This is where WTLS comes into play. It can provide both RSA-based cryptography and elliptic-curve cryptography based encryption and is becoming the de facto standard for security in cellular phones and other small wireless terminals. The requirements for the WTLS protocol are the following:

- Datagram and connection-oriented transport layer protocols must be supported
- Must be able to cope with long round-trip times
- Must recognize that the bandwidth of some users can be low
- Recognize the limited processing power and memory capacity in mobile nodes
- Considerations for the restrictions on using and exporting cryptography

We can observe that WTLS resembles SSL [9] and TLS [7] closely. In fact WTLS is the same as TLS except with suitable modifications to reflect the above requirements for wireless communications.

## 2.3 Wireless Public-Key Infrastructure (WPKI)

Authenticating a message sender's identity is a critical aspect of information security. PKI provides a set of mechanisms based on encryption and digital certificates to achieve authentication, confidentiality, integrity and non-repudiation. The main components of WPKI are:

- End entity (EE) application

- Registration authority (RA)
- Certification authority (CA), and
- PKI directory.

An RA deals with the requests of EE for certificate issues, revocations, and suspensions. Usually, a PKI portal logically acts as a RA. It passes on the requests made by the wireless clients to the RA and CA.

Certificates are attachments to messages issued by a CA, that authenticates a sender's identity and provides encryption keys. The CA uses a public-key cryptography algorithm to generate a public and private key pair. The public key is used to encrypt the message, and the private decrypts it. A sender keeps the private key secure but makes the public key available widely. Any user with access to the public key can encrypt a message and send, while only the user with the private key can decrypt it.

When an EE applies for a certificate, the RA approves the request and sends it to the CA. The CA issues a certificate and posts it in a directory while also sending it to the PKI portal. From here, the location of the certificate is relayed to the EE user. A secure connection is then created between the EE and an application server. Transactions between the EE and the server are then digitally signed and transmitted between them along with the EE's certificate location.

Making PKI work in the wireless world has its own challenges, namely, wireless devices typically have low throughput and computational power, and WPKI systems must interact with their wired counterparts. Therefore, WPKI is in some sense a light weight version of PKI. Just as WTLS is a wireless optimized version of TLS, WPKI is an optimized version of IETF PKIX for the wireless environment. Specifically, optimizations in the following areas have been incorporated [1]:

PKI protocols:

- Certificate format
- Cryptographic algorithms and keys

Wireless PKI for the wireless environment is based on WAP standards. The main components of a WPKI system include the following:

- WAP identity module (WIM) in mobile phones
- WAP gateway with certificate based authentication
- Registration authority
- Back-end PKI infrastructure with access to infrastructure for certification authority

The WIM is a hardware module that contains the security keys and certificates. The hardware module is considered to be tamper-proof.

## 2.4 Smart Cards

Using a subscriber identity module (SIM) card is a popular approach in wireless communications for subscriber authentication and roaming. These cards can store PKI-based authentication information and can be inserted into a wireless device. A smart card has an embedded silicon chip that enables it to store data, communicate via a network, and deliver authentication data. Smart cards are of two types: (a) microprocessor and (b) memory. Memory cards only store data that can be retrieved later by a card reader. Therefore, these are only as secure as the card reader can be. A microprocessor card, on the other hand, can manipulate and edit data in its memory. A third category of smart cards that is emerging is a dual interface card. It features a single chip that enables a contact or contactless interface with a high level of security. When a smart card transmits data wirelessly, then an additional layer of security is essential to protect the data. One method that is used by smart card technologies is scrambling (enciphering) and descrambling. This provides some amount of confidentiality to the transmitted data. Smart cards can do efficient scrambling such that only authenticated cards and computers can communicate with them; making hacking or eavesdropping extremely difficult.

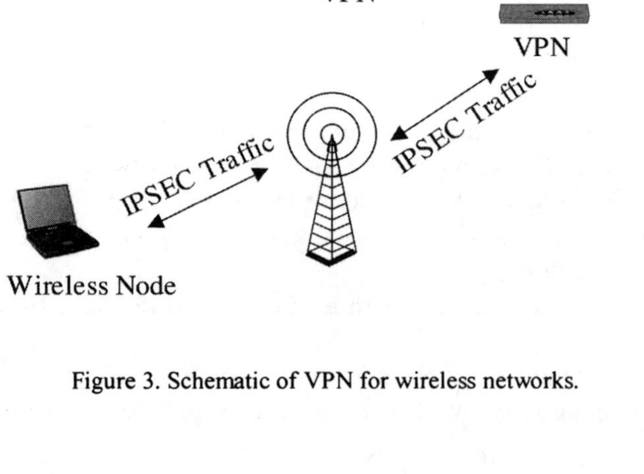


Figure 3. Schematic of VPN for wireless networks.

## 2.5 Virtual Private Networks

The virtual private network (VPN) concept originated from wired networks. For wireless applications, the schematic of a VPN set-up is as shown in Figure 3. Internet-based VPN seem to be evolving into a viable, secure alternative for dedicated leased lines. Since the Internet is not controlled, Internet-based VPNs use encryption to transmit data between VPN sites. With the new standards for network data security on IP networks, it is becoming even more possible to create VPNs. VPNs are expected to provide the following four critical functions to ensure security for transmitted data:

- Authentication: Ensuring that the data originates at the source that it claims to come from
- Access control: Restricting unauthorized users from gaining admission to the network
- Confidentiality: Preventing anyone from reading or copying data as it travels across the Internet
- Data integrity: Ensuring that no one tampers with the data as it travels across the Internet

In VPNs, "virtual" implies that the network is dynamic, with connections set up according to the user needs. It also means that the network is formed logically, regardless of the physical structure of the underlying network (the Internet, here). Unlike leased lines, VPNs do not maintain permanent links between the end points that make up the network. Instead, when a connection between two sites is needed, it is created; when the connection is no longer needed, it is torn down, making the bandwidth and other network resources available for other uses.

**Internet Protocol Security (IPSec)** is the most widely used mechanism for securing VPN traffic. IPSec can use multiple algorithms for encrypting data, keyed hash algorithms for authenticating packets, and digital certificates for validating public keys. VPNs also support several user authentication methods. These standards-compliant methods allow for their easy integration into existing network infrastructures. The IPSec protocol includes three principal security elements:

- *Authentication Header (AH):* The AH provides authentication and integrity by adding authentication information to the IP data. Therefore, it is difficult for an unauthorized user to access or alter the data. Authentication techniques used are MD5 (Message Digest Algorithm 5) and Secure Hashing Algorithm (SHA).

• *Encapsulation Security Payload (ESP):* The ESP provides confidentiality. It can also provide integrity and authentication, depending on the algorithm used. With the ESP in use, part of the ESP header itself and all the data are encrypted. Tunnel and transport modes are available, with tunnel mode being the choice for remote access. Encryption techniques used are Data Encryption Standard (DES) which uses 56-bit keys and Triple-DES or 3DES which uses 168-bit keys.

• *Internet Key Exchange (IKE)*: These are the management protocols that are used to negotiate the cryptographic algorithm choices to be employed by the AH and ESP. The mechanisms used provide for an extremely scalable solution. Keys are maintained, exchanged, and verified using these protocols.

When the number of users is large, such as in a large enterprise network, VPN seems to the best choice for wireless security. It can also be easily managed and controlled.

## 2.6 Other Approaches

In addition to the above discussed approaches, there are several other ways in which to secure a wireless system. We briefly describe in this section some of the techniques proposed by the IEEE802.11 WLAN standard.

### Service Set Identifier (SSID)

The SSID allows a WLAN to be segmented into multiple networks, each with a different identifier. Each of these networks is assigned a unique identifier, which is programmed into one or more access points (APs). To access any of the networks, a client node must be configured with the corresponding SSID identifier for that network. Thus, SSID acts as a simple password identifier. If the SSID is known widely or shared, then this form of security can be weak.

### Media Access Control (MAC) Address Filtering

Each AP can be configured with a list of MAC addresses of the client nodes authorized to access that AP. If a client's MAC address is not on the list, access is denied by that AP. This method provides good security but is not scalable to large networks. It is difficult to manually enter the MAC addresses and maintain them for a large number of nodes.

### Shared Key Authentication

Shared key authentication uses a challenge-response framework. A node willing to authenticate (initiator) sends an authentication request management frame indicating "shared key authentication." The recipient of this request (responder) responds by sending an authentication management frame containing 128 octets of challenge text to the initiator. This challenge text is initiated using WEP peudo-random number generator with a "shared secret" and a random initialization vector. Upon receiving this management frame, the initiator copies the contents of the challenge text into a new management frame. This frame is then encrypted with WEP using the shared secret along with a new initialization vector. The encrypted frame is then sent to the responder. The responder decrypts the received frame and verifies that the 32-bit CRC integrity check value is valid, and the challenge text matches the text it sent originally. If there is a match, then authentication is successful. Following a successful authentication, the initiator and responder swap roles and repeat the same process for mutual authentication.

## 2.7 COMPARISON OF SECURITY APPROACHES

Figure 4 [5] summarizes the user convenience versus security of the different approaches discussed so far in this section. From this figure we see that techniques that minimize human intervention seem to be more convenient and generally more secure. This is quite understandable. However, a number of approaches shown in this figure and discussed in this chapter are vulnerable to attacks. These attacks could be both intentional and unintentional as described in the following section.

# 3. SECURITY VULNERABILITIES AND COUNTERMEASURES

Security flaws in current wireless systems are due to a variety of reasons—some of them due to the technology and the others due to poor system administration and usage. We describe here some of the serious security flaws/attacks that have been discovered recently along with some countermeasures that are being developed.

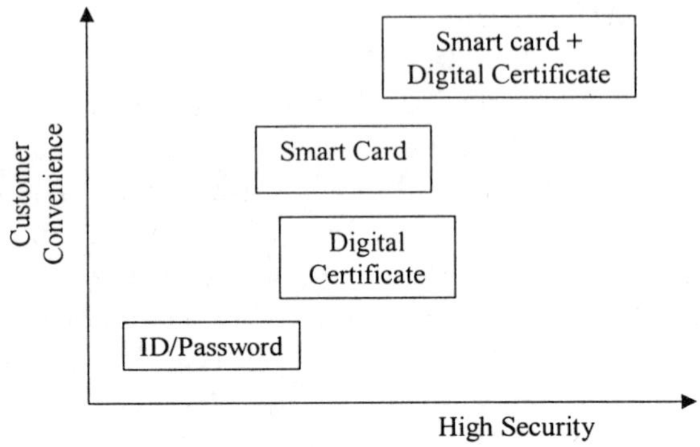


Figure 4. User convenience versus security comparison

## 3.1 Problems in WEP

WEP encryption has been observed to be weak and vulnerable to attacks [6]. Since it lacks support for per-packet integrity protection, a wide variety of attacks, including insertion of packets into the data stream are possible. Therefore alternative ciphers are under development by IEEE 802.11 Task Group I, such as temporal key integrity protocol (TKIP) and wireless robust authenticated protocol (WRAP). Also, since the WEP encryption key is short, a brute force attack can be launched to successfully break it.

The key scheduling algorithm in WEP has also been observed to be flawed. Therefore the secret key can be obtained in time that increases only linearly with the key length. The Task Group I is responsible for investigating remedial measures for correcting flaws in the WEP algorithm. Among the several proposals that were put forward, the obvious one was to increase the key size. A key size of 104-bits, so that it makes a 128-bit seed along with the 24-bit initialization vector was proposed. But, a result due to Fluhrer et. al. [8] showed that the key scheduling itself had problems.

Therefore, even larger keys did not mean more security. RSA Security Inc. [4] has proposed fast packet keying [4] as a solution to this flaw.

TGi is working towards a long term security architecture for 802.11 called the Robust Security Network (RSN). The use of 802.1X standard for Port-based Network Access Control has been proposed. But it has been shown by Mishra et. al. [2] that 802.1X is still vulnerable to session hijacking, man-in-the-middle attacks, and denial of service (DoS) attacks since the authentication is asymmetric (AP is always trusted). It is suggested to use per-packet authenticity for management frames, and symmetric and scalable authentication.

## 3.2 Security Holes in WTLS

A number of potential security problems have been identified in WTLS. As discussed previously, WTLS is a modification of the TLS specification. Some of these modifications seem to be have opened up WTLS to problems such as chosen plaintext attack, datagram truncation attack, message forgery attack, and the key-search shortcut for some exportable keys [11].

WTLS supports unreliable datagram service. It is known that datagrams may be lost, duplicated, or reordered, thereby requiring the cipher block chaining (CBC) mode to use a new initialization vector for encrypting each packet. The used initial vector is computed by XOR'ing the sequence number of the packet and the original initial vector, which is derived during the key generation/linear initial vector computation. The first plaintext block in the packet is then XOR'ed with the computed initial vector. The original initial vector is computed based on values sent during the handshake. All these values containing client-random, server-random and sequence number are sent without encryption, so they can be eavesdropped. Also, predictable initial vectors are vulnerable to chosen plaintext attacks against low-entropy secrets [11].

The WTLS supports a 40-bit XOR MAC (Message Authentication Code) that works by padding the message with zeros, dividing it into 5-byte blocks and XOR'ing these blocks together. For stream ciphers, the XOR MAC does not provide any message integrity protection, regardless of the key length. A bit can be inverted in the ciphertext if the inverting is also done to the MAC.

Thus the integrity check will be successful even when the content has been modified. This security problem also affects integrity.

The record type field is sent without encryption. Therefore, an eavesdropper can determine the change of keys reading the contents of this field. The existence of encrypted error messages can be determined from this field thus causing problems in privacy.

## 3.3 Weakness in MAC Address Filtering

Weaknesses in MAC (Media Access Control) address filtering based security are mainly due to two reasons: (a) MAC address must appear in the clear even with WEP being enabled, and (b) wireless cards permit the changing of their MAC address via software. This means, an attacker can determine the permissible MAC addresses via eavesdropping, and then subsequently gain access to the network.

## 3.4 Flaw in Shared Key Authentication

A flaw in shared key authentication is described in [2]. Assume that an attacker is eavesdropping on a mutual authentication cycle. The attacker first captures the random challenge text transmitted in the clear by the responder to the initiator in the second step of the authentication process (refer to Section 2.6). Then, the attacker also captures the encrypted challenge text (encrypted using the share authentication key) sent by the initiator to the responder as an authentication response. Now, the attacker knows the random challenge text (plaintext), its encrypted version (ciphertext) and the public initialization vector. This enables the attacker to derive the pseudo-random stream produced using WEP, with the shared key, and the initialization vector. The size of this stream will be the size of the authentication frame. Therefore, the attacker knows all the elements of the authentication frame. It is now easy for the attacker to request authentication to join an access point. When the access point responds with an authentication challenge in the clear, the attacker can take the random challenge text and the pseudo-random stream, and compute a valid authentication response frame by XORing these two values. A new integrity check value can be also be computed using the technique discussed in Borisov et. al. [6].

From these discussions we see that attacks on wireless security features range from the most simple to the very sophisticated. If considerable precautions are not taken, it is fairly easy for an intruder to enter a wireless network and eavesdrop without being detected.

# 4. CHALLENGES AND OPEN ISSUES

From the discussions presented so far, we see that there are numerous challenges and open issues in providing wireless network security and privacy. One among the controversies raised by wireless technologies is the FCC mandated, wireless Enhanced (E911) [3] for cellular telephones. E911 seeks to improve the effectiveness and reliability of wireless 911 service by providing 911 dispatchers with location information on wireless 911 callers. This system could locate every cellular phone within an accuracy of 375 feet 67% of the time. Its ultimate goal is to improve the accuracy to 40 feet 90% of the time. While this service has its own benefits, tracking all the cell phone users all the time raises some serious privacy issues.

Denial-of-service attacks in wireless networks are relatively easy. There are cheap devices in the market that can be used to intentionally jam the signals emanating from cell phones, cordless phones, and other wireless communication devices. Detecting the source of jamming and prosecuting those who intentionally jam transmissions is a big challenge.

Security features are known to reduce the wireless throughput, even up to 50% in some cases. Since the throughput is already small (relatively speaking) in wireless networks, how much of it can be lost for additional security protection is a big question. It looks like, a common user is not willing to sacrifice a sizeable chunk of throughput for security.

What are the implications of this? A thorough cost benefit analysis in this regard still seems to be lacking. Users must be allowed to choose the amount of security they are willing to pay for.

Wireless links can go into deep fades thus causing unacceptable bit error rates. It is known that, encryption codes are very sensitive to bit errors. Even a single bit error in the encrypted bit stream could cause havoc. This is known as the "avalanche effect." Most solutions to this issue seem to rely on forward error correcting codes (FEC). FECs can only correct/detect up to a certain maximum number of bit errors. They are also known to exhibit the so-called "brickwall effect", i.e., their performance degrades abruptly after a certain probability of bit error value. The questions that are still open are: (a) is it possible to jointly optimize encryption codes and FECs for fixed wireless channel conditions and security level, and (b) optimal allocation of bit rates for encryption and FEC under a total bit rate constraint.

Intrusion detection in wireless networks is still an open problem. Mechanisms that can reliably differentiate intrusion from interference need to be developed. Quick detection of intrusions is another open research area. Ideas from statistics and probability theory could play a major role in obtaining a solution to this problem.

Since many wireless devices are limited by battery energy constraints, it would be useful to investigate the impact of various wireless security algorithms and protocols on energy consumption. Even though rough estimates for the energy consumption can be derived using theoretical methods, a more realistic approach would be to conduct experimental measurements on real hardware processors with energy and memory capacity constraints. Mathematical models based on the gathered experimental data could reflect the real situation more closely than theoretical derivations. Note that the energy consumption will also vary depending on the software or hardware implementation of the security protocols.

Finally, the security loop holes discovered so far and the ones that will be discovered in the future have to be fixed quickly and satisfactorily.

## ACKNOWLEDGEMENTS

This work was partially supported by grants, NSF ITR CCR-0082064, NSF CAREER ANIR-0133761, and Stevens Center for Wireless Network Security.

## REFERENCES

1. http://www.certicom.com/pdfs/whitepapers/trustpointwireless pki.pdf.
2. http://www.cs.umd.edu/waa/wireless.html .
3. http://www.fcc.gov/911/enhanced/ .
4. http://www.rsasecurity.com .

5. http://www.verisign.com .
6. N. Borisov, I. Goldberg, and D. Wagner. Intercepting mobile communications: The insecurity of 802.11. 7th Annual International Conference on Mobile Computing and Networking, July, 2001
7. T. Dierks and C. Allen. The tls protocol version 1.0 (rfc 2246). http://www.ietf.org/rfc/rfc2246.txt.
8. S.R. Fluhrer, I.Mantin, and A. Shamir. Weaknesses in the key scheduling algorithm of rc4. Selected areas in cryptography, pages 1{24, 2001.
9. A.O. Freier, P. Karlton, and P.C. Kocher. The ssl protocol version 3.0. http://wp.netscape.com/eng/ssl3/ssl-toc.html .
10. http://www.wapforum.org .
11. M. Saarinen. http://www.jyu.fi/»mjos/wtls.pdf
12. S. Weatherspoon. Overview of IEEE 802.11b security.
13. http://www.intel.com/technology/itj/q22000/articles/art 5.htm

**About the Author:**

R. Chandramouli is currently an Assistant Professor in the Department of ECE at Stevens Institute of Technology and serves as a Co-director of MSyNC: Multimedia Systems, Networking, and Communications Laboratory. His research in wireless networking and security, multimedia security, and applied probability theory is currently being funded by the National Science Foundation, Air Force Research Laboratory, NJ Center for Wireless Telecommunications, and Stevens Center for Wireless Network Security, among others. He is a recipient of the NSF CAREER and IEEE Richard E. Merwin Awards. He also serves as an Associate Editor for the IEEE Transactions on Circuits and Systems for Video Technology.

Chapter 8

# SECURING BUSINESS'S FRONT DOOR

*Password, Token, and Biometric Authentication*

Lawrence O'Gorman

*Avaya Labs Research, Basking Ridge, NJ, USA*

**Abstract**: Human authentication is the security task whose job is to limit access to computer networks and physical locations only to those with authorization. This is done by equipping authorized users with passwords or tokens, or using their biometrics. However, due to human limitations, these are often used poorly, thus weakening security, or they are secure but so inconvenient as to be circumvented. This chapter describes common means for authentication as well as their strengths and weaknesses. Some of the major issues are detailed to emphasize the tradeoffs required when considering different authentication schemes. Examples of common systems applications are given with appropriate authentication choices. Finally, future trends are described to help to understand how soon and to what degree the security-convenience tradeoff will be improved.

**Keywords**: password, security token, biometrics, human authentication, user authentication, access control, security.

## 1. INTRODUCTION

### 1.1 Human Authentication

An attempt to identify the most common computer-age frustration needs look no further than the initial authentication stage. Most of us have forgotten the password to our computer at least once if not several times. Many of us have remembered passport, toothbrush, and other items for a trip, only to forget the security token that enables access to the corporate network. And some of us have remembered that security token only to find it useless at the remote machine since it does not have the reader (such as smart card reader) or interface (such as USB) in which to plug the token.

Sometimes there are more serious ramifications of not being able to log in. Consider a situation where a government, military, or medical official must make a decision from a remote location. Since there can be no face-to-face confirmation of identity, how is the person authenticated? A password could be used, but this may be forgotten in an infrequent and stressful situation. A token could be used, but a token such as a smart card requires a reader. The only biometric whose reader is ubiquitous is speaker verification over a telephone line, but this is unreliable if there is a noisy background or when stress alters the voice characteristics. Passwords, tokens, and biometrics are the tools by which secure access can be gained. However, these tools are not equivalent and, unfortunately, none has yet proven to offer perfect security and universal convenience.

Authentication is the process whereby an entity verifies that the claimed identity of another entity is its true and authorized identity. For applications involving computers and telecommunications, this is done for the purpose of performing trusted communications between them. We distinguish between *machine-by-machine authentication* (or simply, *machine authentication*) and *human-by-machine authentication* (or simply, *human authentication*). See Figure 1. Machine authentication is well established and relatively secure. An example is the Secure Sockets Layer (SSL) or Transport Layer Security (TLS) protocol for Internet transactions. When you perform a secure transaction on the Internet, your browser sets up this communications link to protect your data (for instance, your credit card number). This is done transparently to the user by use of secret keys (secret X and secret Y) stored on the respective machines and never seen or known by most users. While this protocol assures that the machines know each other's identity, it gives no assurance of the human identity at the client end. As illustrated in Figure 1, it is often the intention that not just Alice's machine, X, is authenticated, but also that it is indeed *Alice present at client X.* Since anyone gaining access to client X also has access to secret X, there must be human authentication in addition to the machine authentication. This human authentication restricts client X access only to those who know secret A, which in this example is a password belonging to Alice. It is this human authentication that is the subject of this paper.

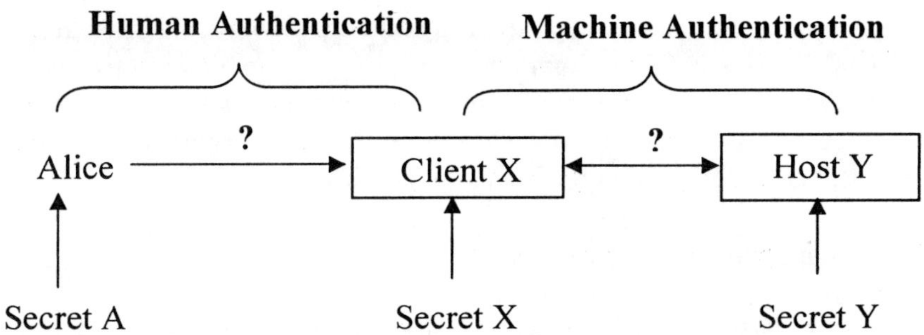


Figure 1. Alice performs human authentication to Client X by demonstrating knowledge of secret A, such as a memorized password. Two machines, client X and host Y perform machine authentication by mutually demonstrating knowledge of their respective stored secrets.

Although human authentication has been practiced far longer than computers and telephones have been in existence, it is much less secure than machine authentication. Human authentication is variously called the "Achilles heel," the "weak link," and the "last yard" of secure systems. Whereas machines have the capabilities to store and remember long secrets, humans do not. Consider that the machine that uses long, random secret keys for machine authentication is protected by a user's password that may be poorly chosen and quickly guessed. The easiest place to attack many systems is via the human authentication process: guessing a password, stealing a token, or spoofing (producing a forged copy of) a biometric.

A listing of security tasks includes the following: authentication, confidentiality, integrity, non-repudiation[1], access control, and availability. For *all* these tasks, authentication can be termed the *gatekeeper*. For instance, a document's confidentiality can be safeguarded by encrypting it. To do this, there is a secret key, which is long and random and cannot easily be memorized by a user, so this is usually stored with access through a memorable password. The integrity of a document can be safeguarded by producing its digital signature that assures it has not been altered. However,

[1] The term, *non-repudiation*, can be confusing. This refers to the ability of a security system to prove that a transaction did take place even though the user involved claims he is not responsible for it. For instance, a user might claim that someone else made a particular credit card charge, so he shouldn't have to pay it. If the system can show it was definitely made by the true user (this is what a handwritten signature is for), then the system is performing non-repudiation.

this again relies on a secret key stored behind a password. Each of the other tasks as well relies on some component assuring access by authorized people and denial to those unauthorized. So, human authentication is the Achilles heel of secure systems in two ways: it is used to secure other security functions, and it depends on humans who are fallible.

## 1.2 Authentication Abuses

Before telephones and computer networks, the traditional mode of authentication was through a human gatekeeper. When a gatekeeper visually recognized the person requesting authentication and knew that he was authorized, the gatekeeper would grant access. However, for remote transactions when personal acquaintance is not the norm, a "shorthand" means to prove authentication is necessary. This can be a password, a physical token, or a biometric. All of these – including the biometric – are only derivative indicators of authenticity; they are not "who you are," but a mutually agreed device or protocol to act as evidence that you are who you say you are. Because these are not who you are, they can be subject to fraudulent use. Statistics show prevalence of this abuse:

- The CERT/CC (Computer Emergency Response Team / Coordination Center), a federally funded organization based at Carnegie Mellon University, estimates that 80% of all network security problems are caused by bad passwords. [1]
- In a study of the FBI Computer Intrusion Squad, 40% of 538 surveyed companies detected system penetration resulting in an average loss of $2M per company per year [2]
- There were 25,000 categorized attempts to break into U.S. government computer systems in 2001 of which 245 were successful [3].
- IDC (an information technology market research company) market projections predict that corporate computer security expenditures will increase from $2.8B in 2000 to $7.7B in 2004, with the fastest growing component being administration, and the second fastest being authentication, $562M in 2000 to $1.7B in 2004, a compound growth rate of 32%.

Secure human authentication is important not only for protecting the corporate (or government) interests, but also for protecting those of the individual. This is because in an electronic or networked application, your identity must be proven to the machine by some artificial means such as a password or identity card. If one of these is stolen, it is as good as your identity being stolen in the machine's view. Here are some statistics of identity theft:

- According to the U.S. Office of the Comptroller of the Currency [4], there were over half a million people affected by identity thefts in the year 2000.
- The identity theft line at the Identify Theft Resource Center of the Federal Trade Commission receives over 3,000 calls per week. Instances of this crime are growing at 30 to 40% per year [5].

## 1.3 The Business of Authentication

Two sentiments sum up both the selling and buying/implementing sides of human authentication:

1. Security is like quality, reliability, and economy. It is less often an end product than it is an important component to successful products and services across a wide range of industries and applications.
2. It is difficult to sell security as a product, but it is much more difficult to sell a product without security.

With respect to point 1, security is an essential component for many business products. For other products, it can provide a "value-added," such as for financial services where a greater number of features or higher value transactions can be offered when stronger security is available. A classical example is the ATM machine, which over the years, offered larger amounts of cash as security offered by the card plus PIN (personal identification number) combination was strengthened. Financial institutions want to derive higher revenues with more services that can only be enabled by better authentication. For these, the end product is not security, but security makes the end product possible.

With respect to point 2, customers are becoming increasingly knowledgeable and demanding in matters of security. As more services are being transacted in a faceless manner across phone and data lines, individuals are more likely to use these services if they have trust in their privacy. They are increasingly aware of their own vulnerability, whether it is from infringement of their private health and financial records, theft of their credit card numbers, or theft of their identities. Of course, there is also a conflict between businesses and services demanding more personal information to protect their own networks versus individuals becoming increasingly concerned about identity theft and their privacy.

## 1.4 Definitions

In this section, we define a few terms used in the rest of the paper.

*Human authentication* is performed to verify that the claimed identity of a person is the true identity and, by implication, that the true identity is authorized for the requested task. Human authentication is also termed *end-user authentication,* or simply, *user authentication*.

We call the entity used to perform authentication an *authenticator*. Passwords, security tokens, and biometrics are all authenticators.

The term *password* is used for passwords, passphrases and PINs. We will use the term, *security token*, or *token,* to include any physical object that facilitates security. This includes an electronic security token, a smart card, and a magnetic stripe card such as an ATM card.

We will often talk about the relative *strength of security* of an authenticator, or simply its security. The security of an authenticator is measured by the difficulty of anyone but the true owner to successfully authenticate with it.

# 2. AUTHENTICATORS

## 2.1 Types of Authenticators

Human authentication factors have traditionally been placed in three categories:

- What you *know*, e.g., password, PIN, mother's maiden name.
- What you *have*, e.g., ATM card, smart card.
- Who you *are*, e.g., fingerprint, iris, or voice biometric.

However, these categories can lead to confusion. A password is not so much what you know, as what you've memorized. A biometric is no more who you are than your height or eye color fully describe you; it's just a physical feature. Instead, we categorize three types of authenticators by how they support authentication (see Table 1):

1. ***Secrets*** support authentication by their secrecy or obscurity. The user and the authenticating entity share a secret that (hopefully) no one else knows. This type includes the memorized password as well as obscure knowledge such as mother's maiden name. A drawback of this category is that, each time the secret is shared for authentication, it becomes less secret. Advantages of a secret include low cost, inexpensive entry via telephones and keyboards, as well as the fact that it can't be physically lost.
2. ***Tokens*** support authentication by their physical possession. The traditional token is the metal key that has stood well the test of time. The drawback of a metal key is that, if lost, it enables its finder to enter the

house. This is why most digital tokens combine another factor, an associated secret PIN to protect a lost or stolen token. There is a distinct advantage of a physical object used as an authenticator, because if lost, the owner has physical evidence of this and can act accordingly.

3. ***IDs*** indicate the identification of a unique owner, and support authentication by their resistance to copying or counterfeiting. A drivers license, passport, credit card, university degree, marriage contract, etc. all belong to this category. So does a biometric, such as a fingerprint, eye scan, voiceprint, or signature. Their dominant security defense is a difficulty to copy. Just as a driver's license should be difficult to forge, a good biometric should be difficult to copy. A disadvantage of these is, if they are compromised (stolen or forged) then they are sometimes time-consuming to replace with a new, changed document – of course, biometrics are impossible to change. An advantage is that, unlike secrets and tokens, a good ID cannot be lent because it is inextricably tied to the true owner by name, photograph, biometric, or any combination of these.

Table 1. Authenticator types, security attributes, examples of each, their main defenses, and main drawbacks.

| Type | Authentication by: | Advantage | Disadvantage | Examples | |
|---|---|---|---|---|---|
| | | | | Traditional | Digital |
| **Secret** | Secrecy or obscurity | Inexpensive | Less secret with each use | Combination lock | Computer Password |
| **Token** | Physical Possession | Evidence of loss | Insecure if lost | Metal key | Car keyless entry |
| **ID** | Uniqueness and Personalization | Cannot be lent | Difficult to replace | Drivers license | Biometric |

Authenticator types can be combined to reap benefits in security or convenience or both. This is commonly called, *multi-factor authentication*. A common multi-factor authenticator is an ATM card, which combines a token with a secret (PIN). If a user has difficulty remembering the secret, a token may be combined with a biometric. The photo-ID is the traditional 2-factor ID plus biometric. Rarely is a secret combined with a biometric ID, since the objective of the biometric is usually to eliminate the task of memorizing the secret. To date, there has not been much application for 3-factor authentication, at least for business applications.

## 2.2 Secrets

There are many studies showing the vulnerabilities of password-based authentication schemes [6 - 11]. The basic problem with passwords can be explained succinctly: a memorable one can often be guessed or searched by an attacker and a long, random, changing password is difficult to remember.

There are a number of ways that technologists have tried to mitigate the memorization problem. Some of these will be addressed under future trends in Section 5. However, one answer that is currently available and gaining popularity is single sign-on. Single sign-on is a remote service or client-side software package that stores multiple passwords. One password is required to access this intermediary, and then subsequent password-access is handled transparently (without additional passwords required from the user). One example is corporate single sign-on, to which an employee will gain password access, then all machines within the corporation to which the employee is authorized are available without requiring more passwords. Single sign-on service is also available for the Internet. Here, the user stores login information to all her Internet sites at the service, through which access is enabled. These 3rd party services are more convenient than remembering multiple passwords, but they only solve the password problem within their own domain, corporate or Internet, so multiple passwords must still be remembered. Single sign-on can also be accomplished at the user end by use of a smart card (or some other token) storing multiple passwords. The user authenticates herself to the smart card with a single password and stored passwords are transmitted transparently. There is also an option akin to single sign-on. This is a software package on the user's machine or PDA (personal digital assistant), which stores encrypted passwords. A single password decrypts any stored password, which is then entered manually by the user for access. There are pros and cons of these options that are discussed in Section 3.

## 2.3 Tokens

A security token is a physical device that can be thought of as portable storage for authenticator(s). This can be a bankcard, a smart card containing passwords, or an active device that yields time-changing or challenge-based passwords. Since these devices are designed for security applications, most are secure to physical and electronic tampering. ATM magnetic stripe cards are at the lower end of security and cryptographically-enabled smart cards are at the high end (see FIPS 140-2 on standards for high-end security requirements [12]). In this chapter, all security tokens are two-factor authenticators with password (or biometric) access to the token. The token can store human-chosen passwords, but an advantage is to use these devices to store longer codewords that a human cannot remember. We use the term, *codeword,* to be a secret number like a password, except it is machine-generated and machine-stored, so it can be longer, more random, and perhaps changing.

There are two categories of tokens:

1. Passive or Stored Value, e.g., bank card
2. Active, e.g., one-time password generator.

The passive storage device is usually a magnetic stripe or smart card in which a static codeword is stored. The active device usually contains a processor that computes a one-time password, either by time-synchronization[2] or challenge-response. Some active tokens can also perform cryptographic calculations to encrypt and decrypt. A smart card can participate in challenge-response protocols from the authenticating server by virtue of a cryptographic processor on the card. This provides strong security, but requires a smart card reader at the client. Note that only higher end "secure" smart cards have a cryptographic processor. If they do, we categorize them as active tokens; if they only store passwords or codewords, but can do no processing, they are passive devices.

The primary advantage of tokens is their security. Because they can store or generate a codeword much longer than that which a human can remember, they are much less easily attacked. A token that yields a 12-digit codeword has $10^{12}$ possible different codewords. This is called its *keyspace* and it is advantageous that the keyspace be as large as possible to combat guessing and brute force attacks. Compare this to a password. An eight-character password that fully and randomly uses the full 62 ASCII alphanumeric characters actually has a keyspace of over $10^{14}$. However most users are not usually so diligent in their password selection and average password keyspace is around $10^6$. Therefore, in the average case a token is more secure than a password as measured by keyspace,

Token ($10^{12}$) >= Password ($10^{14}$-$10^6$).

---

[2] A widely-used, proprietary, active device is RSA Security's *SecurID*™ token. This produces a different, randomly chosen codeword each minute and is synchronized to the server, which receives the password. Since the user enters the codeword displayed on the token, it has the advantage of not requiring special readers at the client.

Another advantage of a token is that it provides compromise detection. If a password is stolen and even being used, the legitimate user might have no idea. However, if a token is stolen, its owner has evidence of this fact by its absence. A final advantage is that it narrows the opportunity for denial of service attacks. An attacker must steal the token –requiring physical presence – which will effectively cause denial of service to the token owner. If only a password is used, an attacker can remotely enter incorrect passwords until the system freezes the account, thus causing denial of service.

Three downsides to security tokens are:
1. The user must remember to physically possess it to authenticate.
2. Most security tokens require the user to memorize a PIN, so this effectively adds the memorization drawbacks as for passwords.
3. Most tokens require a port or reader to convey information to the machine, for instance a smart card reader, USB port, etc. These may not be widely available or available across different access modes such as computers and telephones.

## 2.4 Biometric Ids

A *biometric* is a feature measured from the human body that is distinguishing enough to be used for user authentication. Biometrics include: fingerprints, eye (iris and retina), face, hand, voice, and signature, as well as some other more obscure or futuristic biometrics [13, 14] such as gait and smell. A biometric offers to inextricably link the authenticator to its owner, something passwords and tokens cannot do, since they can be lent or stolen. When used to verify the person involved in a transaction, this inextricable link can offer *non-repudiation*, proof that a transaction did take place involving the authenticated user though that user denies it. However, biometrics are not without issues. We examine some of these in Section 3.

Biometrics are traditionally classified as follows:
- *Physical biometric*, including fingerprint, iris, retina, face, hand geometry, etc.
- *Behavioral biometric*, including voice, and handwriting.

The physical type includes biometrics based on stable body features. The behavioral type includes learned movements such as handwritten signature, keyboard dynamics (typing), and gait. Speech is usually categorized as behavioral because it is a product of learned behavior; however the underlying body feature upon which speech biometrics is based is the vocal tract, which is physical and relatively stable. Of course, one can also argue that gait, etc. are also related to stable body features.

Due to these ambiguities inherent in the traditional classification, we prefer a different one. Instead of classifying the physical body characteristic itself, we classify the *signal* that we measure – a signal that is based on the characteristic but may or may not be a direct measure of it. We define two types:

1. Stable biometric signal.
2. Alterable biometric signal.

A *stable biometric signal* has the property that the pertinent information used to match and differentiate is a biometric template that derives directly from the fixed body characteristic. For example, a fingerprint yields a stable biometric signal because there is no change in the signal from the body features to the signal used for matching.

An *alterable biometric signal* has two components. One is the underlying body characteristic, which should be fairly stable such as to serve as a good measure for authentication. The other is a variable that alters the biometric. For instance, speech produces an alterable biometric signal because a person's stable vocal tract is used to verbalize a variable phrase. Similarly for handwriting, a user's fairly stable handwriting style is used to write a chosen signature or password. Furthermore a biometric such as face that can generate a stable biometric signal can also generate an alterable biometric signal if the face is measured voicing a particular sentence or reacting to a particular stimulus.

A fingerprint is a stable biometric signal that is matched on the pattern and features of the ridges that constitute the fingerprint (Figure 2). Fingerprints used to be the tool only of criminal investigators, but fingerprint readers have evolved to small and inexpensive microelectronic chips, so they

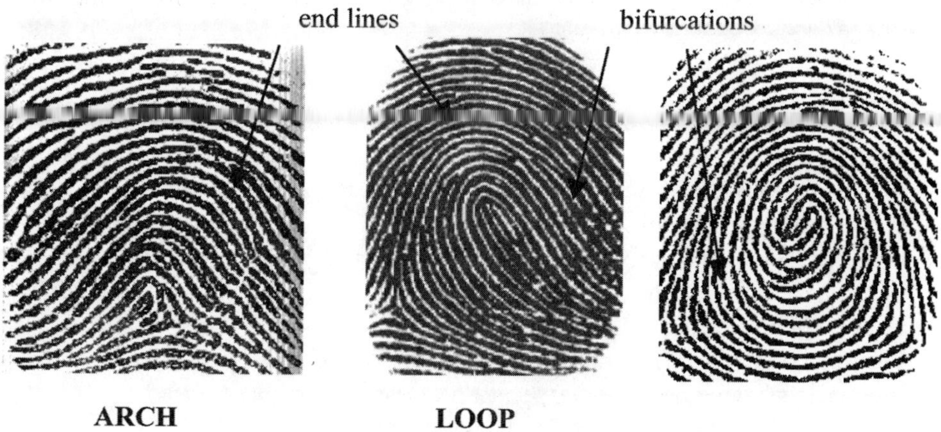


Figure 2. Three patterns of fingerprints. Besides these patterns, fingerprints are matched based on locations of end lines and bifurcations, as shown

are now practical for use in door entry systems, laptop computers, and even car key fobs [15, 16]. An advantage of fingerprint systems is their ease of use. A disadvantage is their variability with dirty, sweaty, or dry skin, sometimes causing false rejections of the legitimate user.

Eye-based biometric systems produce stable biometric signals. There are two approaches to eye biometrics (Figure 3). The iris biometric uses an image of the iris (the color area surrounding the pupil). The iris can be captured non-invasively by a camera that looks no different than one to photograph your face. For the retina, a user is required to look into a lens tube, whereupon a laser light scans the retina through the pupil to produce an image of the arterial pattern there. Because of the more invasive nature of this, retinal scanning has not gained much acceptance beyond prison exit/entry stations. The advantage of both of these eye biometrics is their very high accuracy in matching users. A disadvantage has been cost, traditionally much higher than for fingerprints, however the cost is decreasing making implementation economical at least for public door access or ATM machine verification.

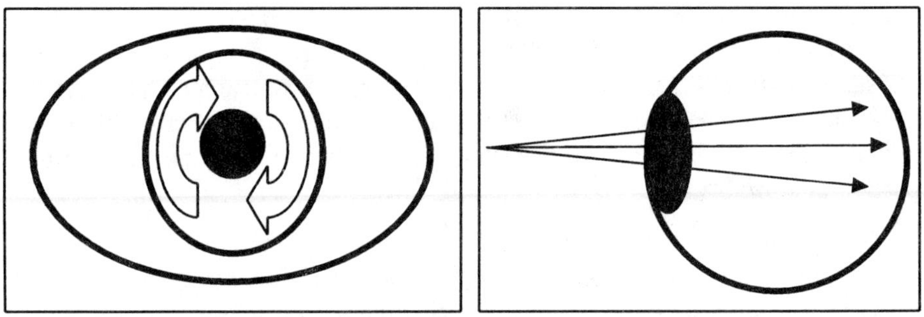

Figure 3. Eye-based biometrics. On the left, the iris area is shown from which iris biometric systems extract matching features. On the right, light is shone through the pupil to image the back of the eyeball for retina systems.

Face biometric systems derive their stable biometric signals from facial features (Figure 4). An advantage of these systems is that the face can be imaged even with no required cooperation of the user. However, computer facial recognition has not produced as high accuracy as most other biometrics, so is rarely used alone for secure verification applications. Instead, facial recognition systems manufacturers have directed their efforts at surveillance, such as finding criminals in airports or other public areas. For this application, facial systems have an advantage over most other biometric systems because cooperation of the user is not required.

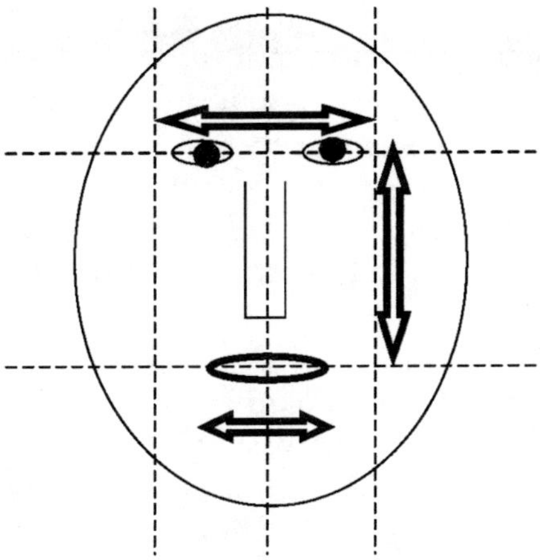

Figure 4. Facial features such as inter-eye distance and distance from the eye line to the mouth are used for some face biometric systems.

Hand geometry produces a stable biometric signal that includes finger lengths, heights of knuckles, distances between joints, etc. (Figure 5). Although less well known than fingerprint, eye, and face systems, hand geometry systems have had a longer history of use (outside of law enforcement) than the others. Because a reader must be as large as a hand, they are not appropriate for installing in laptops, nor have they been accurate enough for very high security tasks, but they are accurate and rugged enough to have been implemented for use in health clubs and university cafeterias as well as for many other applications.

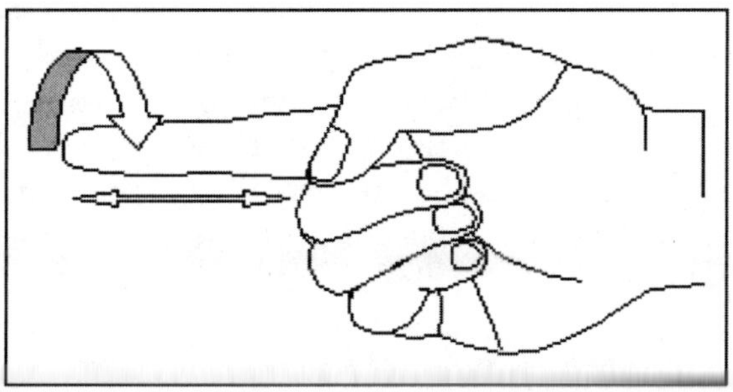

Figure 5. Hand geometry systems measure features of the hand, such as finger lengths and knuckle heights.

Voice biometric systems, or speaker verification systems, measure the signal characteristics of a vocalized phrase (Figure 6). A vocalized phrase is an alterable biometric signal because the same speaker can vocalize different phrases. There is a distinct advantage of biometric systems that rely on alterable biometric signals. If a fingerprint, or iris, or face is "stolen" – or compromised – the legitimate user cannot easily change it and will have to abandon that biometric. However, if a spoken pass phrase is stolen such as by audio recording, compromise recovery is as simple as changing the pass phrase. The downside of speaker verification is that, although relatively accurate under ideal conditions, background noise and variability in a user's voice (such as a result of laryngitis) prevent consistently high recognition results.

There are other biometrics – keyboard dynamics, handwriting, palm, ear, smell, facial thermograph (pattern of veins under skin in face), gait, etc. – however those described above are the ones most widely considered at this time.

## 3. CONSIDERATIONS WHEN CHOOSING AUTHENTICATORS

There is no perfect authenticator. Below, some considerations are described of particular authenticators. In Section 4, appropriate choices are given for some systems taking into account these considerations.

### 3.1 What's Wrong With the Status Quo?

The password would actually be a pretty good authentication factor – if only human capabilities could meet technological demand. We defined keyspace in Section 2.3. Below is the keyspace comparison of passwords versus tokens and biometrics. (The effective keyspace for biometrics can be estimated as the inverse of the false accept rate (see [17]) using 3rd party recognition rate data shown in Table A-1 in the appendix.)

Token ($10^{12}$) >= Password ($10^{14}$-$10^{6}$) > Iris ($10^{6}$) > Fingerprint, PIN ($10^{4}$) > Voice ($10^{3}$) > Face ($10$-$10^{2}$).

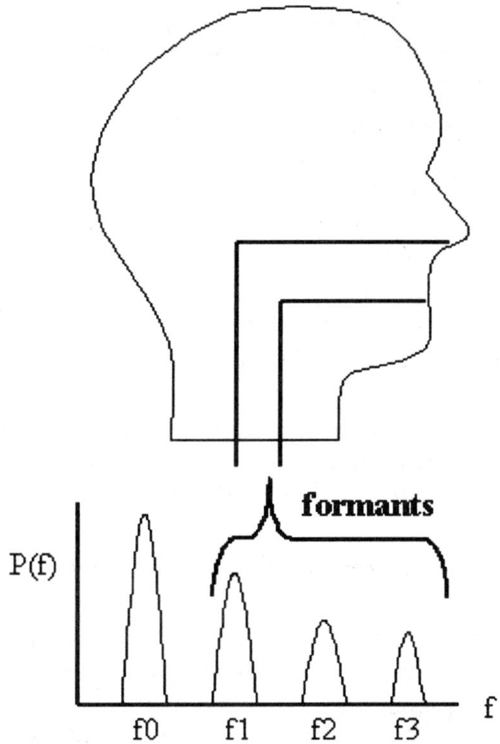


Figure 6. Distinctive features for verification can be measured from spectral frequency characteristics of a voiced pass phrase.

One can see that a well-chosen password can be even stronger than a token (with 12-digits) and much better than a fingerprint. The problem with passwords is largely human. Strong ones are difficult to remember. Multiple passwords exacerbate this difficulty. And, when these must be changed regularly, this goes beyond the patience if not mental capabilities of most humans.

Obscure knowledge used for authentication (such as mother's maiden name) should, by definition, be only narrowly known and difficult to find. Multiple pieces of information are often required to further narrow down the potential knowledge group. This approach is more convenient, but there are several security drawbacks. One is that it is difficult to measure the obscurity of one or more pieces of information. Are questions A, B, and C good enough? One person's obscure knowledge may be another person's public information, in which case more or different questions are needed. Another problem is that these depend in part on the scheme's own obscurity: if mother's maiden name were the information needed to gain access to everyone's bank account, it is likely this information will not be obscure for long. Yet another problem is when an obscure knowledge system asks the user to choose an obscure question. For instance, if the user chooses a question asking about his favorite breakfast cereal, the user might think this pretty clever – how would an attacker know this? However, the smart attacker would just guess cereals by popularity and likely come upon the user's cereal in a few guesses. Putting the onus of strong security upon the common user is likely to fail. A final problem is related to compromise recovery. You recover from a compromised password by changing it. What do you do after your mother's maiden name is compromised?

Having stated many problems with passwords and obscure knowledge authentication approaches, there are researchers working on methods to improve these. See Section 5 to see how the status quo might be improved.

## 3.2 Rejecting Customers is Not Good for Business

In the introduction to Section 3, we stated that there is no perfect authenticator. Let's examine this from the perspective of rejecting customers. If the customer has a password, there is the onus on her to remember it. When she forgets it, she may feel one of a number of emotions, none of them that a particular business would want associated with it. She may feel that her mental capacity is not up to snuff. She may feel aggravated with the security requirement. She may feel she is wasting her time to have to go to the password reset mechanism, which may either be a person or an obscure knowledge system. She may think about aborting the transaction or taking her business elsewhere.

A token may or may not be easier for the customer. A common token is a magnetic stripe card, such as a credit card, frequent flier card, etc. Another common token is a key fob card inscribed with bar code such as is used at many grocery stores. There are RFID (radio frequency identification) contactless cards and wands as well that can just be passed in front of a reader to transfer the customer information[3]. There are also smart cards (where it is necessary that the customer has a smart card reader wherever the smart card must be used[4]). One can understand from the variety of form

[3] The Exxon Mobil SpeedPass device is the first RFID device to have gained widespread popularity for retail purchases.

[4] American Express introduced the "Blue" card in the late 1990s, in part to perform secure on-line transactions with a smart card. With this card was the necessity that customers also

factors offered that the issuer must decide on the easiest device such that the customer will allocate wallet, purse, pocket, or computer space to have the token when needed. There should also be a procedure in place such that customers can make transactions even after misplacing the token.

A biometric does not tax the customer's memory. However, biometrics are sometimes rejected despite the best intentions of authorized users. This is a situation arguably worse than forgetting a password or misplacing a token. A customer may be doing everything right, but be rejected. This is frustrating for the customer, and thus not good for the company that has deployed the unreliable biometric system.

What are the chances that customers will be rejected? Using the 3rd-party recognition results from Table A-1 in the appendix, even the best biometrics will produce rejected customers. Take an example of the 70,000 airline passengers that travel through Newark airport daily. If a good fingerprint system rejects 1% of these people, there will be 700 angry customers per day. Using even the best results on the table of 0.25% rejection error for an iris system, there will be 175 rejected customers per day. Many people will accept this understanding the security requirements of an airport, but if this happened at a grocery store, they might be less tolerant. Although any individual customer should not be so worried for himself – because each has a low chance of rejection – the business has ample cause for concern because it is rejecting many good customers per day.

In practice, if a biometric system usually works for a particular customer, it will rarely reject that customer. For those who are incompatible with the technology for whatever reason, the onus is on the business to provide an alternative. Therefore, effective authentication systems that include a biometric will also offer options, usually just the password.

## 3.3 Falsely Accusing a Customer is Even Worse

Surveillance is a function that is being undertaken at some public sites. The objective is opposite the use of authenticators for access, it is the use of biometrics for identifying and denying access. Face recognition is particularly well suited among biometrics for this task because it requires no cooperation from the user. Perhaps few businesses would have use for this, however casinos for example have deployed these systems to identify those

have a smart card reader. This was provided free of charge to good customers as a loyalty incentive.

they would prefer not in their establishment. A business must be careful when it is identifying someone they'd prefer *not* to do business with. If rejecting a good customer is bad for business, misidentifying someone as a bad customer is even worse.

Consider the following example. A casino has a database of 25 people from throughout the world who they would prefer not play their high-stakes game. Of course, they would prefer as many other high rollers as possible play this game. Once each week 300 people enter the room – past a face recognition camera – to play the game. Those whose faces match any of the 25 in the database are rejected. What is the rate of false match? For this example, let's assume a face recognition system has been set to a most liberal security setting to falsely reject as few as possible. From Table A-1 in the appendix, choose a false match rate of 0.001, or 1 in 1000. One might guess that, with this rate, only one person will be falsely identified every 3.3 weeks, but this is untrue. The true mis-identification rate is 7 in 300. How does a false match rate of 1 in 1000 yield 7 misidentified in 300? The reason is due to the database of 25 people. If there was only one person in the database, the guess above would be correct. With 25 people in the database, the false match rate is,

$$\mathrm{FMR}(n) = 1 - [1\text{-}\mathrm{FMR}(1)]^{n}$$
$$\mathrm{FMR}(25) = 1 - [1 - 0.001]^{25} = 0.0247.$$

This is about 25 times worse than for a database of one, and this makes sense since there is 25 times the number of faces to mismatch. Now, for 300 people per week, the total number likely to be misidentified is,

$$\begin{aligned}\mathrm{Total}(300) &= 300 \times \mathrm{FMR}(25)\\ &= 300 \times 0.0247 = 7.4.\end{aligned}$$

So, over 7 people per week will likely be misidentified as those you *don't* want to do business with. This is a substantial number of good customers to reject.

## 3.4 When "Design for Failure" is "Revert to the Status Quo"

One of the tenets in designing security systems is to "design for failure." That is, if an attacker compromises the system, there should be layers in addition to the authentication layer that detect and stop this attack. Intrusion detection systems are important for this very reason. Implicit in this design is that no single security feature is 100% effective. For a password system, if a user finds his account compromised, he can change the password through a password-reset service. If a token is lost or stolen, it can be canceled and replaced with a different one. However, biometrics cannot be replaced. If you find your fingerprint is being used without your knowledge, someone is likely using a fake, or "spoof," rendition of your fingerprint. You might be able to detect this, but what can you do about it? You can't change your fingerprint. The same is true for your face, your iris, your hand, or your voice. Furthermore, you can't keep most biometrics secret – without wearing gloves to leave no fingerprints, wearing a veil to hide face and eyes, or ceasing all speech to thwart sound recording. Unfortunately, the "design for failure" plan for most biometrics is "revert to the status quo," which means going back to passwords [18].

At this time it is relatively difficult to copy biometrics without knowledge of the user, although it can be done more easily with the user's cooperation. Biometrics designers are developing more advanced copy detection methods. This is like the counterfeiting business: the authorities develop anti-counterfeiting features in a currency, which the attackers eventually beat, whereupon the authorities design stronger features ... and this cycle continues. The one big difference between this and biometrics is that human body features cannot be improved with more anti-copy features, so when attackers have reached close to perfection in copying there is nothing to do but abandon the biometric. When a business is considering adopting biometrics, it must be considered whether security savings will recoup infrastructure costs before the technology is widely vulnerable to copying.

The problems for biometrics described above suggest "the paradox of secure biometrics." This paradox applies only to stable biometric signals, and is described below:

1. A biometric is said to be unique. This is good.
2. However, something unique can never be changed. This is not so good for compromise recovery.
3. Furthermore, a biometric is not a secret [19], so it can be found and potentially copied. If so, it is not unique. This is bad.
4. So, is a biometric not a good authenticator after all?

## 3.5 Different Authenticators Are Different

The response to the paradox of Section 3.4 avoids the question of whether uniqueness and biometrics are good or bad. Instead, it points to the mistake of judging different authenticators all by the same criteria. Different authenticators have different features, both advantageous and disadvantageous. For instance, passwords can be kept secret and changed when compromised; biometrics cannot. However, biometrics can be convenient and they are not easily lent, contrary to passwords. A biometric is not a password or token replacement, but neither are a password or token a biometrics replacement.

Table 2 shows some of the major strengths and weaknesses of different authenticators including their combinations. Some explanations are required. For secrets, "infrastructure in place" refers to the fact that these need only a keyboard or telephone in contrast to a special purpose reader needed for most tokens and biometrics. An exception is the voice alterable biometric signal because the telephone is a ubiquitous reader. Compromise detection for a token refers to the fact that a token is a physical object whose absence can be seen, whereas if a password or biometric is stolen there is no indication of this until after the fact of its illicit use. Non-repudiation is listed as a strength of biometrics even though we have said there is the possibility of an attacker copying a biometric. This is because it has stronger ability to support non-repudiation than a password or token that can easily be lent or lost. Challenge-response is a security protocol that protects against the common replay attack. Virtually all strong authentication procedures employ challenge-response. Since a stable biometric signal is fixed, it can only respond in one way to a challenge; therefore it cannot support a challenge-response protocol. An alterable biometric signal can support this.

Which of the authenticators from Table 2 is best? It depends upon how much security is desired, how important convenience is to the users, and the cost of adoption and maintenance. Different authenticators and their combinations will be appropriate for different requirements.

## 3.6 Single Sign-On Service – Can You Trust it?

Although single sign-on services promise to mitigate password memorization problems, they only reduce the number of passwords that must be memorized. Furthermore, if the one password to a service is guessed or stolen, then an attacker can gain access to all sites or machines served by that service. Therefore, it is even more critical to choose a strong password (of sufficient length, with random characters, and not in the dictionary) than otherwise.

Table 2. Authenticators compared by strengths and weaknesses.

| Authenticator Types | | Strengths | Weaknesses |
|---|---|---|---|
| **Secret** | Password/PIN | Inexpensive<br>High (potential) keyspace<br>Infrastructure in place | Can be forgotten<br>Can be lent<br>Poor non-repudiation<br>Cost of password reset |
| | Obscure knowledge | More easily remembered<br>Inexpensive<br>Infrastructure in place | Vulnerable to guessing |
| **Token** | Static | Compromise detection<br>High keyspace | Can be lost/stolen<br>Reader required |
| | Active | Challenge-response secure<br>High keyspace | Cost |
| **ID** | Stable biometric signal | Convenient<br>Better non-repudiation<br>No challenge-response | Reader required<br>Compromise recovery<br>Recognition errors |
| | Alterable biometric signal | Challenge-response secure<br>Ubiquitous reader (phone)<br>Better non-repudiation | Recognition errors |
| **Combination** | Token + PIN | Strong 2-factor security<br>Challenge-response secure<br>Compromise detection | Can forget token or PIN<br>Reader required |
| | Token + biometric | Strong 2-factor security | Cost for 2 readers<br>Recognition errors |

For an Internet single sign-on service or any 3rd-party service, a prospective user should consider whether it is trustworthy or not. This is because all your Internet sites, all your purchase details, all your interests, and all your passwords may be available to the single sign-on service. Even if you trust the company to be honorable, are the privacy measures adequate to prevent their administrators from learning your information? It is very difficult to make anything perfectly secure or private. A single point of attack is much more convenient for an attacker to learn all about you than if the information is distributed at multiple sites and behind multiple passwords. Of course, this is the tradeoff between convenience and security.

## 3.7 Smart Cards and Tokens – Can You Trust Yourself?

A single sign-on alternative to a service is a smart card or other secure token that stores all passwords, PINs, secret keys, digital certificates, etc. This has several advantages over a single sign-on service:

- You need not trust anyone else with your personal information, since you hold and maintain the token.
- The token is portable, not requiring network connection as a service requires.
- It is usually off-line, reducing vulnerability to network attacks.
- Physical evidence of a lost token provides compromise detection.

However, there is at least one major disadvantage of your own single sign-on token – maintaining it. Since loss of the token would mean loss of all passwords, etc., it is imperative to back it up. For security concerns just mentioned, the backup should not remain on-line. Maintaining security of the card via software improvements and patches is all the job of the cardholder. You may not trust a service with all your personal information, but can you trust yourself to diligently perform maintenance and upkeep?

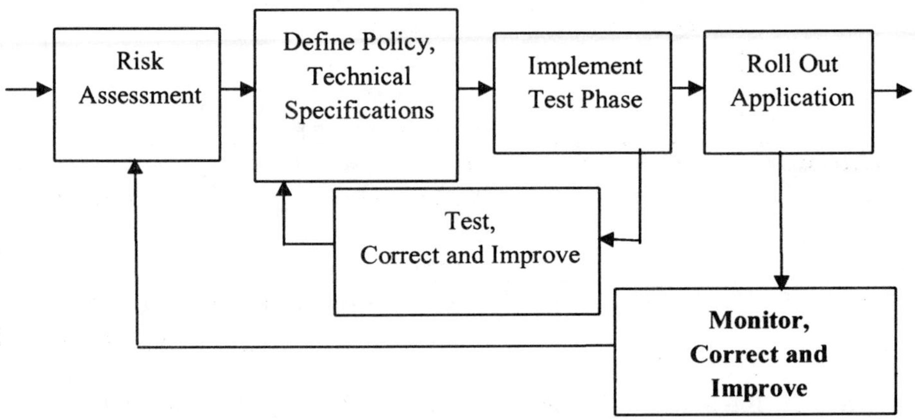


Figure 7. General procedure for building a security system.

## 4. SPECIFYING AUTHENTICATION SYSTEMS

It must be emphasized that authentication system design does not stop at the choice of an authenticator. More important than the component parts is holistic design of the security system. Part of this is the necessity to monitor, correct and improve – since we have stated that security systems can fail. Just as attackers improve their methods and try different attacks, a good security system must also be improved. A diagram showing secure system design tasks is given in Figure 7, with the emphasis on continually designing, monitoring, and improving.

Keeping in mind that the proper authenticator choice is only part of the complete system solution, we discuss below appropriate authenticator choices for particular applications.

### 4.1 Network Access

The password has been the standard for computer network access for decades. If the system employs some secure challenge-response password transmission protocol [20, 21] and limits the number of failed authentication attempts (as most systems should), it will be resistant to most attacks. Since passwords can be lent, this choice does not offer non-repudiation. Nor does it offer compromise detection. Password maintenance is straightforward, however it may be costly when passwords are forgotten, especially if system policy mandates good, non-dictionary passwords and frequent changes. A commonly quoted cost for each instance of password reset is $30-$50. The problem with a password-only system is that people either forget their password, incurring maintenance costs, or they choose a memorable password, which might also be guessable and that weakens the security of the system.

A password plus token combination is the more secure choice for authenticating network access. The penalty is an increased system cost for the token, reader, and system software. There is a convenience cost for the user as well because she still has to remember a password for the token and also has to remember to carry the token.

### 4.2 Physical Access

In many corporate applications, secure physical access (such as to a building or to a computer room) can employ the same types of authenticators as for network access. In some cases, especially where public safety is concerned such as at airports, there needs to be stronger assurance that the person possessing the authenticator is its true owner. Since passwords and tokens can be guessed, lent or stolen, this is an application for biometrics. Because of the concern that biometrics can be copied – either now or in the future – a biometric should always be paired with something else, most commonly a token. If this is a smart card token, the biometric can be matched directly to the user's biometric data on the card, then a codeword is generated on the card and sent to the server for authentication. This protects the privacy of the user's biometric template data.

## 4.3 Non-Network Access

It is more difficult to protect a non-network access (such as opening a password protected file that is on the user's machine) than a network access. One reason is the inability to limit the number of authentication attempts since there is no administrator to reset the password. The other is the fact that a password is usually stored locally with the password-protected file and susceptible to attack.

The straightforward password solution is fine if the user has enough discipline to create and remember a long, random password, and to choose different passwords for different protected files. Another solution is a token that can securely store many passwords. This has additional advantages that it can store multiple passwords for different applications, it is non-networked to limit network attacks, and it facilitates compromise detection.

## 4.4 Telephone Access

One of the drawbacks of some of the authentication schemes mentioned is the need for a client-side reader for token or biometric. Readers may become more widely installed, smart card readers being the most likely, however there is already a biometric reader that is ubiquitous. This is the telephone that can be used for speaker verification. Call center applications involving health care or financial records are well suited for this authentication method. Unlike stable biometric signals, speaker verification can be an alterable biometric signal with the advantage described in Section 3.5 of supporting a challenge-response protocol for added security. In fact, a voice biometric has several advantages over other biometrics (Table 2), with the only, but major, downside being that the current recognition rate may not be adequate for many secure applications.

## 4.5 Access by Identification

The ultimate convenience would be for a customer to carry no token, remember no password, and not even enter a name, but let the system identify him as one out of many. Biometric identification is the only authenticator mode that could do this, but the problem is challenging. Consider a system whose database numbers 10,000 customers. We'll use the best false match rate of one in a million for iris recognition (from Table A-1 in the appendix). The system false match rate is,

$$\text{FMR}(10{,}000) = 1.0 - (1.0 - 10^{-6})^{10{,}000} = 0.00995.$$

If 1,000 transactions are made per week using this identification system, there will likely be 1000 x 0.00995 = 9.95 false matches per week. This means that about ten people per week will receive bills not attributable to them. Most businesses would not abide this.

There are two instances where an identification scheme like this would be usable. One is where the database is much smaller. For access to a room or small company where the number in the database is in the 10s, this is feasible. The other is where customer convenience and high tech pizzazz outweighs finances. Consider a theme park where season ticket holders are admitted simply using a biometric. The security setting can be set such that there are few false rejections at the cost of relatively high false matches. The legitimate customer is happy because she is rarely rejected and the park is much less concerned with the few attackers who are admitted with a false match.

For most applications, it is likely not such a burden to ask the customer to supply some identifying piece of information. Then the challenge of one-to-many identification is reduced to an easier task, one-to-one verification or one-to-few identification.

# 5. FUTURE TRENDS

Despite the choices available for human authentication, there is still much improvement to be made. The two top issues with current authentication schemes are:

1. Most schemes suffer a security-convenience imbalance – if they are convenient they are not secure enough, and if they are secure enough they are inconvenient.

2. Most strong schemes cannot be used widely because special readers are not currently available.

Corporate and university labs are attempting to remedy these issues. Some project examples are outlined below.

**Graphical passwords** are claimed to be more memorable to users. The *Déjà vu* project at the University of California at Berkeley [22] displays an array of abstract images to a user, who chooses the ones she has memorized. The *HumanAut* project at Carnegie Mellon University [23] requires the user to choose the pictures he has memorized from a sequence of memorized and other pictures. The *Draw-a-Secret* project at Bell Labs, AT&T Labs, and NYU [24] requires the user to make a line drawing in the same shape and sequence within an invisible grid pattern.

**Enhanced tokens** include multi-function smart cards that store multiple passwords on a single token and can perform other tasks, such as employee identification (employee badge) or cafeteria debit. For wireless convenience, new security tokens will contain an RFID or Bluetooth chip, both for wireless detection in the proximity of a reader. This will contribute to a concept called *presence*, where machines can sense when you are close to them with no action from you. PDAs will also be enhanced with hardware and software to securely store passwords and other secure or private information.

**New and Multi-modal biometrics** attempt to address some of the shortcomings of current biometric solutions. Multi-modal biometrics combine different biometric modalities to strengthen security, reduce false rejections, and provide alternatives to the user. New biometrics include gait recognition, infrared capture of blood vessel patterns, and implantable chips. See [25] for links to these and other biometrics.

**Personal Q&A** schemes are advancing past the stereotypical mother's maiden name. Often, users are invited to create their own questions and answers that are most memorable to them. Many corporate security systems for password reset use specific knowledge held by that system. For instance, an airline might ask about recent flights, and a brokerage might ask about mutual funds owned. Researchers are working to improve these schemes to quantify the amount of security the system can expect from particular questions (akin to keyspace) [26, 27], and to guide users to good (high security and memorable) questions and answers. The terms "obscure knowledge" and "out of wallet" are also used to describe this approach.

These allude to the fact that the information is obscure, but not secret, something that may appear on cards carried within one's wallet or purse.

## 6. SUMMARY

Human authentication is a critical concern for corporate security because it is the first layer of defense in protecting company resources and because it is (arguably) the weakest layer due to the human factor. The problem originates with passwords, the status quo of authentication, which would provide good security if only humans could remember strong passwords. Tokens, including smart cards and active devices, can provide stronger security, but these put a burden on the human to remember to carry the device. Biometrics address the convenience issues, but there are unanswered security issues such as what to do if a biometric is compromised – you can't change your fingerprint. Single sign-on services are now being offered to reduce the memorization load, but this is at the expense of trusting a single entity to provide strong security and respect your privacy. Although there is no single perfect solution for human authentication, we have attempted to provide insights into advantages and disadvantages of the current options, what is appropriate for different applications, and what is upcoming from current research.

## APPENDIX

To compare the biometrics with other authenticators, we have used four sources of benchmark data from respected, 3rd-party sources (unaffiliated with biometric vendors). These are listed below by: name; main sponsoring organization; date of testing; biometric type; some test descriptors; test population size; and reference:

**NIST Speaker '99**; NIST; Mar.-Apr. 1999; voice; telephone quality, text independent, up to 1 minute duration; 233 target trial speakers, 529 imposter trial speakers (test "1-Speaker Detection"); [28].

**FRVT 2000** (Facial Recognition Vendor Test); DARPA; Mar.-Jun. 2000; face; mugshot pose, ambient probe lighting, mugshot gallery lighting, time separation 11-13 months (test "13"); 467 probe faces, 227 gallery faces; [29].

**FVC 2000** (Fingerprint Verification Competition); University of Bologna; Jun.-Aug 2000; fingerprint; 500dpi, 256x364 size capacitive sensor (test "DB2"); 100 fingerprints; [30].

**CESG Biometric Testing Report**; CESG; May-Dec. 2000; face, fingerprint, hand, iris, vein, and voice; standard verification mode of operation for each system, failure-to-enroll separated, time separation 1-2 months; about 200 subjects; [31].

From these data, we have chosen operating points described by false match rate and false non-match rate, (*FMR, FNMR*), error rate pairs that apply to some practical situations for display in the table. Any particular application must choose its own operating point depending upon its requirements and use, and must do preliminary testing (and continuing testing as per Figure 7) to confirm and improve results.

*Table A-3.* Recognition point pairs chosen from results of benchmark testing for several biometric types and different tests.

| Biometric | Test | Test Parameter | Attempts | FNMR | FMR |
|---|---|---|---|---|---|
| **Face** | FRVT | 11-13 mo. spaced | 1 | 16% | 16% |
| | CESG | 1-3 mo. spaced | 3 | 6% | 6% |
| **Fingerprint** | FVC | Mainly age 20-30 | 1 | 2% | 0.02% |
| | CESG | Mainly age 30+ | 3 | 2% | 0.01% |
| **Hand** | CESG | - | 1 | 3% | 0.3% |
| | CESG | - | 3 | 1% | 0.15% |
| **Iris** | CESG | - | 1 | 2% | 0.0001% |
| | CESG | - | 3 | 0.25% | 0.0001% |
| **Voice** | NIST | Text independent | 1 | 7% | 7% |
| | CESG | Text dependent | 3 | 2% | 0.03% |

## REFERENCES


1. SecurityStats.com, "Password Security", http://www.securitystats.com/tools/password.asp
2. Computer Security Institute, "Cyber crime bleeds U.S.," 7-Apr-01, http://www.gocsi.com/press/20020407.html
3. D. S. Onley, "DISA official: users should be accountable for security," *Government Computer News*, 25-Apr-01, http://www.gcn.com/vol1_no1/daily-updates/4028-1.html
4. U.S. Office of the Comptroller of the Currency, Comptroller of the Currency Administrator of National Banks, NR 2001- 41, 30-Apr-2001.
5. Federal Trade Commission, Identity Theft Resource Center, http://www.consumer.gov/idtheft/
6. R. Morris, K. Thompson, "Password security: A case history," *Comm. ACM*, Vol. 22, no. 11, Nov. 1979, pp. 594-597.

7. B. L. Riddle, M. S. Miron, J. A. Semo, "Passwords in use in a university timesharing environment," *Computers and Security*, Vol. 8, no. 7, 1989, pp. 569-579.
8. D. L. Jobusch, A. E. Oldehoeft, "A survey of password mechanisms: Weaknesses and potential improvements," *Computers and Security*, Vol. 8, no. 8, 1989, pp. 675-689.
9. D.C. Feldmeier and P.R. Karn, "UNIX password security – ten years later," *Advances in Cryptology – CRYPTO '89 Proceedings*, Springer-Verlag, 1990, pp. 44-63.
10. J. Bunnell, J. Podd, R. Henderson, R. Napier, J. Kennedy-Moffat, "Cognitive, associative, and conventional passwords: Recall and guessing rates," *Computers and Security*, Vol. 16, no. 7, 1997, pp. 645-657.
11. S. M. Furnell, P. S. Dowland, H. M. Illingworth, P. L. Reynolds, "Authentication and supervision: A survey of user attitudes," *Computers and Security*, Vol. 19, no. 6, 2000, pp. 529-539.
12. National Institute of Standards and Technology, U.S. Department of Commerce, "Security requirements for cryptographic modules," FIPS 140-2, May 2002.
13. A. Jain, R. Bolle, S. Pankanti (ed.s), *Biometrics: Personal Identification in Networked Society*, Kluwer Press, The Netherlands, Nov. 1998.
14. S. Pankanti, R. M. Bolle, A. Jain, "Biometrics: The future of identification," special issue of *Computer*, Vol. 33, no. 2, Feb. 2000.
15. X. Xia, L. O'Gorman, "Innovations in fingerprint capture devices," special issues of *Pattern Recognition*, Vol. 36, No. 2, 2003.
16. L. O'Gorman, "Practical Systems for Personal Fingerprint Authentication," IEEE Computer, Feb. 2000, pp. 58-60.
17. L. O'Gorman, "Seven issues with human authentication technologies," *IEEE Workshop on Automatic Identification Advanced Technologies*, Tarrytown, NY, Mar. 2002, pp. 185-186.
18. B. Schneier, "Biometrics: truths and fictions," *Crypto-gram*, 15-Aug-98 http://www.counterpane.com/crypto-gram-9808.html
19. S. M. Matyas Jr., J. Stapleton, "A biometric standard for information management and security," *Computers and Security*, Vol. 19, no. 5, 2000, pp. 428-441.
20. S. M. Bellovin, M. Merritt, "Encrypted key exchange: Password-based protocols secure against dictionary attacks," *Proc. 1992 IEEE Computer Society Conference on Research in Security and Privacy*, 1992, pp. 72-84.
21. Thomas Wu. "The secure remote password protocol," *Proc. of the 1998 Internet Society Network and Distributed System Security Symposium*,

pp. 97-111, San Diego, CA, March 1998. http://citeseer.nj.nec.com/article/wu98secure.html

22. R. Dhamija and A. Perrig, Déjà Vu: A User Study Using Images for Authentication, *9th Usenix Security Symposium*, August 2000.
23. Nicholas J. Hopper and Manuel Blum, " Secure Human Identification Protocols", In: *Advances in Crypotology, Proceedings of Asiacrypt 2001.*
24. Ian Jermyn, Alain Mayer, Fabian Monrose, Michael K. Reiter, and Aviel D. Rubin. *The Design and Analysis of Graphical Passwords.* In Proceedings of the 8th USENIX Security Symposium, August, Washington DC, 1999.
25. Biometrics Consortium, www.biometrics.org
26. N. Frykholm and A. Juels. Error-Tolerant Password Recovery. In P. Samarati, ed., *Eighth ACM Conference on Computer and Communications Security*, pages 1-8. ACM Press. 2001.
27. C. Ellison, C. Hall, R. Milbert, B. Schneier, "Protecting secret keys with personal entropy," *J. of Future Generation Computer Systems*, 16 (4), Feb. 2000, pp. 311-318.
28. A. Martin, M. Przybocki, "The NIST 1999 Speaker Recognition Evaluation-An Overview," *Digital Signal Processing*, Volume 10, Numbers 1-3, 2000, pp. 1-18, http://www.idealibray.com/links/toc/dspr/10/1/0
29. D.M. Blackburn, J.M. Bone, and P.J. Phillips, "FRVT 2000 Evaluation Report," Feb. 2001, www.dodcounterdrug.com/facialrecognition/DLs/FRVT_2000.pdf
30. D. Maio, D. Maltoni, R. Cappelli, J.L. Wayman, A.K. Jain, "FVC2000: Fingerprint Verification Competition," *IEEE Trans. Pattern Analysis and Machine Intelligence*, to be published, 2001. http://bias.csr.unibo.it/fvc2000/Downloads/fvc2000_report.pdf
31. T. Mansfield, G. Kelly, D. Chandler, J. Kane, "Biometric product testing final report," *CESG report*, March, 2001, www.cesg.gov.uk/technology/biometrics

**About the Author:**

Lawrence O'Gorman is a Distinguished Member of Technical Staff at Avaya Labs Research where he works in areas of security and signal processing. Before this he was Chief Scientist of Veridicom, and prior to that he was a Distinguished Member of Technical Staff at Bell Laboratories.

Dr. O'Gorman has written over 50 technical papers, several book chapters, and three books. He has over 15 patents, and is a contributor to four biometrics and security standards. He is a Fellow of the IEEE and of the International Association for Pattern Recognition. In 1996, he won an R&D 100 Award and the Best Industrial Paper Award at the International Conference for Pattern Recognition. He is on the Editorial Boards of four journals and a member of several technical committees. He received the B.A.Sc., M.S., and Ph.D. degrees all in electrical engineering from the University of Ottawa, University of Washington, and Carnegie Mellon University respectively.

Chapter 9

# MANAGING INFORMATION SECURITY RISKS ACROSS THE ENTERPRISE

Audrey Dorofee
*Carnegie Mellon University, Software Engineering Institute*

**Abstract:** There are many ways to approach information security. This chapter looks at information security from a risk management point of view by focusing on the evaluation of information security risks. The evaluation looks at information-related assets that are critical to the success and survivability of an enterprise. By looking at these critical assets, determining how they are at risk, and developing mitigation plans and protection strategies, the enterprise can take a strategic approach to securing vital information. Taking a broad look at all types of threats, enterprises can spend their limited resources wisely, without falling down the rat hole of a technology-centric solution.

**Key words:** information security risk management, risk evaluation, security risks, security threats, critical information assets, security practices

## 1. A RISK-BASED APPROACH TO INFORMATION SECURITY

For a long time, information security was the domain of the information technology (IT) department. Security concerns rarely impinged on other departments or corporate management. The tools, techniques, methods, and processes were IT-centric, not business related. All of this is changing as survivability of critical information becomes one of the cornerstones of enterprise success. Information security must be acknowledged and established as a legitimate, ongoing business process. This chapter looks at information security from a business risk management point of view within the context of a method developed at the Software Engineering Institute (SEI).

There are many ways to approach information security. At the SEI, within the Networked Systems Survivability program, we have taken a risk-based approach to looking at information security across an enterprise. Managing risks is already a part of the corporate world. Managing information security risks is just an extension of that; an information security risk is just another type of organizational risk. Figure 1 illustrates a simple model for information security risk management. Information security risks are identified, analyzed, and mitigated. However, they must be considered in the context of the business goals, management and organizational practices, and the current laws and regulations.

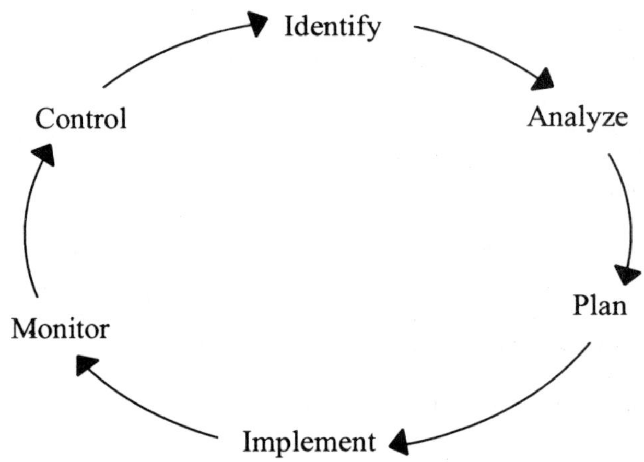


Figure 1. Information security risk management.

One of the more common methods of establishing a baseline set of enterprise risks is risk evaluation. An evaluation should establish a snapshot of the current set of information security risks within the context of the business goals and drivers. We began our research at the SEI by looking at those information security evaluations that existed, different types of risk evaluations, and different ways of modeling security threats.

The first result of our work was the Operationally Critical Threat, Asset, and Vulnerability Evaluation (OCTAVE$^{SM}$)—a self-directed, asset-based, risk-driven information security evaluation, the first part of an information security management process (see Figure 2).

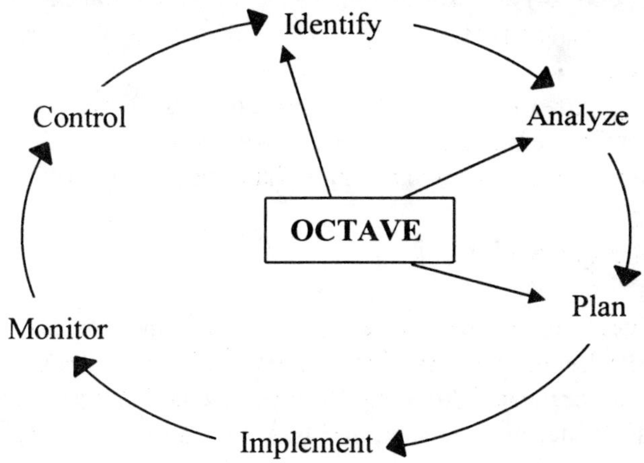


Figure 2. Risk evaluation.

## 1.1 What Problems Are We Trying To Solve?

The security evaluation products and services that existed in 1999 varied widely. They were usually

- reactive (to an event), focused on the past and on what went wrong
- examinations of the computing (technology) infrastructure, which can miss practice-based threats to key assets
- carried out by an external consultant without participation from the enterprise; if the enterprise is not involved in finding the threats or determining the recommended fixes, they don't feel the need to make the corrections; it isn't "theirs"
- producing results that were framed in a standard definition of priority levels, rarely in terms of the customer's business drivers; there was no consistent means to measure the results against an enterprise's business objectives to prioritize threats and improvements.

Typically, the enterprise did not follow through on the results by implementing the recommended changes. In addition, an effective use of scarce resources requires looking towards the future—being proactive.

Something else was needed to address all these shortcomings. The focus of the security evaluations needed to be expanded to be

- both organizational and information technology focused, bridging the gap between management and information technology
- proactive rather than reactive; a risk-based approach

- based on an enterprise's unique risk factors, their tolerance for risk, what is important for their survivability, and the enterprise's business goals and objectives
- inclusive of security policy, practices, procedures, and regulations
- a foundation for continuous security improvement
- a means for institutional learning about information security.

## 1.2 A Risk Perspective

A risk perspective provides an enterprise with a more proactive, forward-looking view. Rather than waiting for vulnerabilities to be found or problems to occur and then repairing them, the enterprise could determine where they were most vulnerable, most at risk, and then decide on cost-effective means of recognizing, resisting, and recovering from the most critical threats. Another difference between technology-focused evaluations and risk-based evaluations like OCTAVE is the use of critical, information-related assets to drive the scope and direction of the process, rather than the technological systems. Technology vulnerability evaluations can then be focused on where the critical assets are located or accessed.

An organization should also evaluate itself against a set of known security practices to get a broad view of how vulnerable it is as a whole. Evaluating the current state of practices against a known standard or collection of security practices helps determine what security practices are relevant, what needs to be improved, and the relative priority of those improvements, based on the risks to critical assets. Without a clear picture of current security practices, risk mitigation actions could be ineffective. Without a risk perspective, practice improvements could be unfocused.

It is easy to forget that these risks, and information about them, exist at all levels and parts of the enterprise and relegate risks to the IT department. *Effective* information security risk management requires cooperation and input from everyone, a partnership among

- all levels of staff,
- business units and the IT department,
- partners,
- contractors,
- service providers, and
- end users.

Without this partnership, critical information about the risks is never discovered and the mitigating actions put in place may fail to correct the problem. Just as the information about risks is distributed, so is the responsibility for securing the assets.

Another aspect of effectively managing risks is making your own decisions, not relying on decisions made by others. OCTAVE uses self-direction to achieve this. Self-direction enables organizational learning and builds a foundation for internal security improvements. Self-direction is managing and directing your own process, leveraging your enterprise's expertise, making your own decisions, and only using external help if needed. Enterprises that completely outsource risk evaluations often detach from the process. They provide little contextual information to the evaluators and do not have insight into the underlying thought processes or algorithms used in the evaluation. The responsibility has been shifted to the external expert, who is not accountable. The enterprise will not understand the underlying risk assumptions or the different possibilities that might unfold. The improvements may never occur or become institutionalized. If an evaluation is outsourced, the enterprise must play an active part in both providing knowledge and making decisions, building a solid, relevant foundation for the future.

## 1.3 Practical Considerations

There were several practical considerations that drove our development of an alternative evaluation method. These considerations apply to nearly every enterprise.

- You can neither eliminate nor mitigate all risks. We all live with some degree of "accepted risk"; you cannot eliminate all risks and still live reasonably.
- Your budget is not limitless. Neither are your other resources.
- You cannot prevent all determined, skilled incursions. However, you can be prepared for them.
- You need to make the best use of your limited resources to ensure the survivability of your enterprise.

We all make decisions based on what we have available and what is most important. This field is no different.

# 2. OCTAVE APPROACH

All of these considerations and issues went into the development of OCTAVE. The roots of information security risks are the information-related assets critical to achieving the enterprise's mission. An effective evaluation focuses on assets and security practices to address the most important threats and risks.

One key aspect of OCTAVE is expanding the view of information security across the enterprise and making the communication links between IT and the rest of the enterprise. As more of the success of the enterprise rests on its capacity for efficient, adequate security, the need grows for good communication and shared vision between IT and enterprise managers. The enterprise-wide focus of OCTAVE is achieved through the involvement of senior and operational managers, staff, mission areas, information technology, etc. Their participation creates the opportunities to share information, viewpoints, and concerns about what is important to protect and how secure the enterprise really is.

Recognizing that there can be many ways to accomplish this type of information security risk evaluation, we distilled the essential elements of OCTAVE into a set of requirements that any good evaluation method should meet. The OCTAVE criteria [1] define a risk-driven, asset focused, enterprise-wide, self-directed, information security risk evaluation. As seen in Figure 3, any number of methods can meet the OCTAVE criteria. Other third parties are now developing their own versions of OCTAVE, tailored to their own specific domain. The SEI has developed two: the OCTAVE Method [2] for large enterprises and OCTAVE-S [3] for small enterprises.

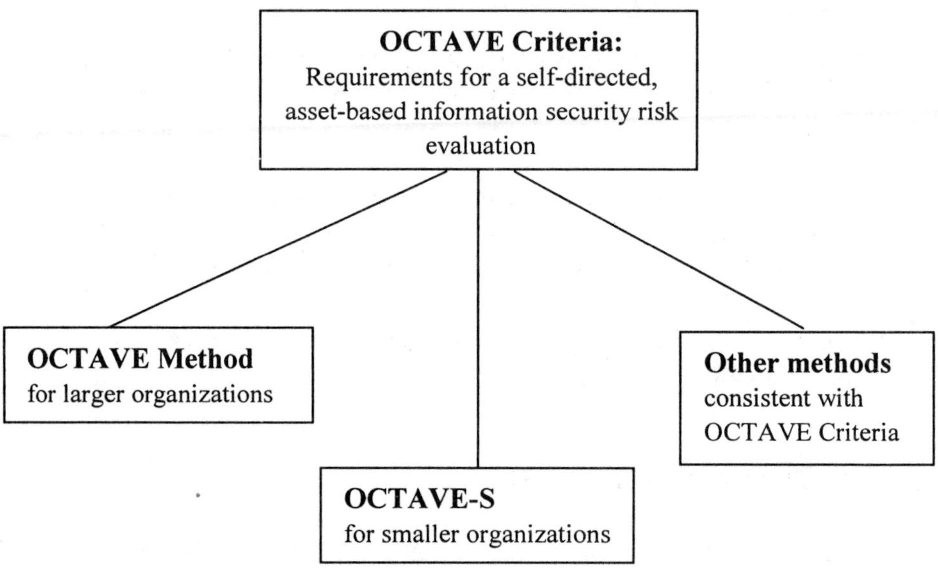


Figure 3. The OCTAVE Approach

## 2.1 OCTAVE Criteria

The criteria include a set of principles [1] – the fundamental concepts that define the philosophy behind OCTAVE. These principles should be considered by any enterprise looking for a security evaluation method. The principles are listed below.

- *Self- direction* – people in an enterprise manage and direct the information security risk evaluations. They make the decisions about the enterprise's security efforts.
- *Adaptable measures* against which an enterprise can be evaluated are essential. A flexible process can adapt to changing technology and advancements. It is not constrained by a rigid model of current sources of threats or by what practices are currently accepted as "best."
- Using a *defined process* and standardized evaluation procedures can help to institutionalize the process, ensuring some level of consistency in the application of the evaluation.
- *Foundation for a continuous process* – an enterprise must improve its security posture over time, making good security practices part of the way it routinely conducts business.
- A *forward-looking view* requires an enterprise's personnel to look beyond the current problems by focusing on risks, potential problems, to the enterprise's most critical assets.
- Every enterprise faces constraints on the number of staff members and funding that can be used for information security activities. By *focusing on the critical few*, enterprises can apply those scarce resources to the most critical information security issues.
- *Integrated management* requires that security policies and strategies be consistent with organizational policies and strategies, striking a balance between business and security goals.
- Information security risk management cannot succeed without *open communication* of security-related issues. Information security risks cannot be addressed if they aren't communicated to and understood by the enterprise's decision makers.
- With a *global perspective* the enterprise creates a common view of what is most important, consolidating individual perspectives.
- No single individual can understand all of the security issues facing an enterprise. Information security risk management requires an interdisciplinary approach, *teamwork*, including both business and information technology perspectives.

## 2.2 Same Criteria, Different Methods

The OCTAVE Method is focused on large-scale organizations (300 and above staff members). It is a systematic, context-sensitive method for evaluating risks using a series of workshops conducted by an interdisciplinary analysis team. In this method, a cross-section of people is selected to provide organizational knowledge in data gathering workshops, including senior managers, operational area managers, staff, and information technology staff. A large enterprise will usually be able to conduct the focused technology vulnerability evaluation either themselves or with an existing third-party contractor. At the end of the evaluation, an additional workshop is held to report findings and recommendations to senior managers, who make the final decision on what actions to take.

OCTAVE-S is designed for the unique needs of the small (20-100 people) organizations. In smaller organizations, it is easier to find an interdisciplinary analysis team with a very broad knowledge about the enterprise and what is important, eliminating the need for data gathering workshops. However, it not as easy to find people with security expertise and IT is frequently outsourced. Security practices may be close to non-existent. The technology vulnerability evaluation may not be possible. Instead, a discussion can be held of the key system components supporting the critical assets and how well they might be able to protect those assets. In OCTAVE-S, the analysis team frequently includes senior managers who can make the final decisions on what actions to take for risks to critical assets. In early pilots, the analysis teams included the chief information officer, the chief financial officer, and even the president and vice-president.

The OCTAVE Method is very open-ended; the questions leave a lot of room to maneuver when developing answers. OCTAVE-S is far more structured, providing a reasonable set of pre-defined answers that can be accepted, changed, or rejected. It uses "fill-in-the-blank" as opposed to "essay" style. A considerable amount of security-related information is embedded in the OCTAVE-S worksheets. Both methods lead to similar results, but they take a different path to get there. A team of people with a lot of security expertise might find OCTAVE-S too constraining while a team with little security expertise would be overwhelmed by the choices they faced with the OCTAVE Method. The next section uses a more detailed look at the OCTAVE Approach to highlight the core aspects of a good information security risk evaluation.

## 3. A CLOSER LOOK AT OCTAVE

When evaluating information security risks, you need to collect information about the enterprise, investigating the technology infrastructure, and determining the appropriate measures for improving security. OCTAVE has three phases, shown in Figure 4. These phases focus on the following three aspects of the evaluation.

- Phase 1 takes an organizational view, identifying the critical information-related assets of the enterprise, their security requirements, the threats to those assets, and the current organizational strengths and weaknesses relative to a set of known security practices.

- Phase 2 is a technology view, similar to the vulnerability tools and evaluations usually conducted on a system. This phase uses the information from Phase 1 to focus the technology view to those key components that support the critical assets.

- Phase 3 takes all of the information from the first 2 phases and identifies the critical risks to the enterprise's critical assets. The enterprise-wide protection strategy and asset-specific mitigation plans are developed during this phase.

OCTAVE is a focused evaluation; the broad-based view of the enterprise is achieved by carefully defining the scope of the evaluation on a broad range of operational areas or business units and using an interdisciplinary analysis team to bring alternative viewpoints. This focused evaluation looks at key aspects, using scarce resources to drive out critical information, and enabling a simpler gap analysis to evaluate any remaining aspects. The cross-section of key operational areas of the enterprise should include those units critical to the overall business goals or survival of the enterprise and always includes IT. For example, a hospital may chose surgery, a lab, patient administration, and outpatient care, plus their information technology department. A manufacturer may select administration, sales processing, the shop floor, and information technology. The focus of the evaluation can expanded or contracted by changing the number of operational areas, or even sites, that participate, or to include partners, vendors, customers, or contractors.

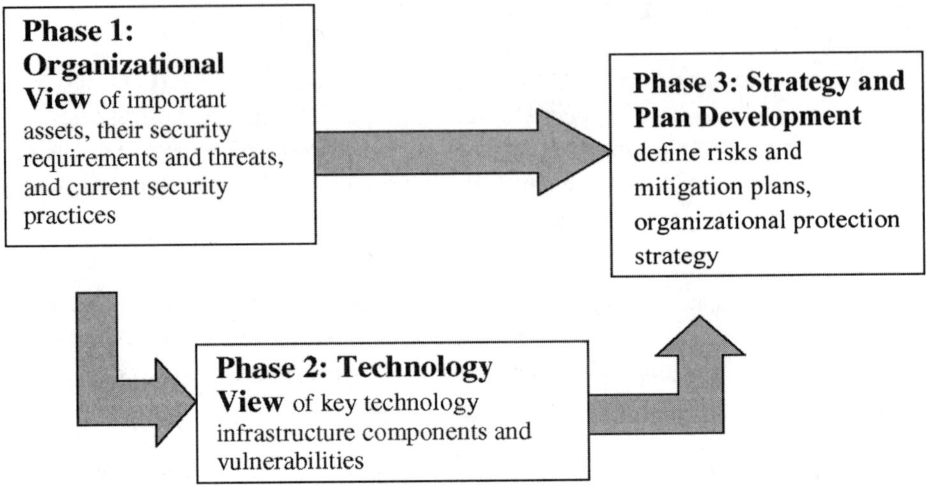


Figure 4. OCTAVE phases.

In OCTAVE, an interdisciplinary analysis team of 3–5 personnel facilitates the process and analyzes the data. By selecting people from across the enterprise, different viewpoints are included in the data analysis and selection of mitigation activities. This team can be made up of business or mission-related staff, information technology staff, strategic planners, or key policy stakeholders. It can even include representatives from contractors, vendors, or partners for even broader viewpoints. The analysis team runs the processes, analyzes the information, and makes the recommendations for an overall protection strategy and mitigation plans.

To understand more about what this type of risk evaluation would include, we'll look at some of the important activities and concepts of OCTAVE, starting with Phase 1.

## 3.1 Phase 1: Organizational View

Phase 1 is focused on the enterprise itself, to answer these questions:

- What are your enterprise's critical information-related assets?
- What is important about each critical asset?
- Who or what threatens each critical asset?
- What is your enterprise currently doing to protect its critical assets?
- What weaknesses in policy and practice currently exist in your enterprise?

**What are your enterprise's critical information-related assets? What is important about each critical asset?**

OCTAVE is oriented around assets – the information or information-related property of an enterprise. OCTAVE focuses on the assets that are vital to the enterprise's mission and business goals. These assets can fall into the following categories:

- information – documented (paper or electronic) information or intellectual property
- systems – information systems that process and store information, including software and hardware components
- applications – software (operating systems, database applications, custom applications, etc.) that processes, stores, or transmits information
- people – a set of people, including their skills, training, knowledge, and experience

Critical assets could be patient and billing information in a hospital, product designs at a manufacturer, or customer information at a service provider. The key is to focus on information-related assets. Intangible assets such as reputation actually come into play later, when you consider the impacts of threats on the enterprise.

**Who or what threatens each critical asset?**

Once critical assets are identified, the threats to those assets need to be investigated. A threat is an indication of a potential undesirable event. One way to consider threats is to develop threat scenarios based on a generic set of threat sources or a generic threat profile. A threat profile contains a range of threat scenarios for a critical asset. OCTAVE uses the following sources of threats:

- human actors using network access
- human actors using physical access
- system problems (for systems under your control)
- other problems (problems due to conditions out of your control)

A possible scenario is a disgruntled employee deliberately using network access to view online personnel records and learning personnel information about managers or a virus interrupting staff members' access to a customer database.

In OCTAVE, the threat profile is visually represented using asset-based threat trees, one for each of the four sources of threats. A common structure to organize threat information provides the framework for discussing and deciding appropriate courses of action to deal with threats.

Each threat has the following specific properties:

- asset – something of value to the enterprise
- actor – who or what may violate the security of an asset
- motive (optional) – defines whether the actor's intentions are deliberate or accidental
- access (optional) – how the asset is accessed by the actor (network access, physical access)
- outcome – the immediate result of violating the security of an asset

Figure 5 shows the human actors using network access tree. The solid lines are the perceived threats. For example, the upper line shows there is a threat of accidental interruption of service for the asset from an employee, using network access. The threats on this tree result from deliberate or accidental actions by people both inside and outside the enterprise. *Inside* threats are people that you consider to be part of your enterprise (e.g., a staff member accidentally deletes an important file while accessing the system). *Outside* threats are people that you do *not* consider to be part of your enterprise (e.g., a competitor uses a network attack to view your enterprise's customer database). You might find some ambiguity in how you think about certain people. For example, a contractor who is resident at your site might be considered an insider.

A similar tree is built for human actors using physical access. For example, could an employee walk out the door with an unencrypted laptop containing confidential data? The asset-based threat tree for system problems would show issues related to systems configured and maintained by your enterprise. The asset-based threat tree for other problems is for conditions that are out of your control. This generally includes threats because of your dependence on other enterprises, natural disasters, and anything not covered by the other categories. For example, if your electrical service was interrupted for an extended period of time, could your assets be disclosed or destroyed?

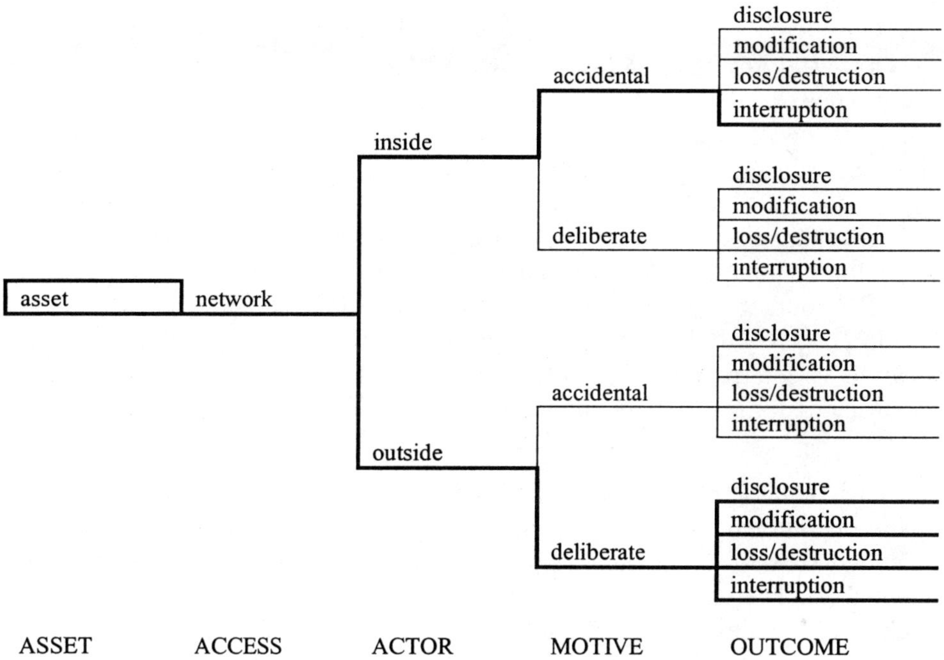


Figure 5. Human Actors Using Network Access Tree

Generic threat profiles can be tailored to suit a particular domain. For example, an organization dealing with national security might prefer an expanded set of threat actors, such as the following:

- non-malicious employees
- disgruntled employees
- attackers
- spies
- terrorists
- competitors
- criminals
- vandals

Trees could also be decomposed into further levels of detail. For example, the network access tress could be broken into different trees for different systems or sets of components. In any case, the threat profile used as a basis for defining threats to an enterprise's critical asset should be relevant to that organization and its domain.

**What is your enterprise currently doing to protect its critical assets? What weaknesses in policy and practice currently exist in your enterprise?**

When deciding what you do well (current security practices) and what you don't do well (current organizational vulnerabilities), you compare yourself against a set of known or accepted security practices. The OCTAVE catalog of practices [4] is one such set, as are the Health Insurance Privacy and Accountability Act (HIPAA) security rules [5] and BSI7799 (ISO17799) [6]. The OCTAVE catalog is a general catalog; it is not specific to any domain, enterprise, or set of regulations. It can be modified to suit a particular domain's standard of due care or set of regulations.

The catalog of practices is divided into two types of practices: strategic and operational, shown in Figure 6. Strategic practices focus on enterprise-level issues at the policy level and provide good, general management practices. They address business-related issues as well as issues that require enterprise-wide plans and participation. Operational practices focus on technology-related issues related to how people use, interact with, and protect technology. Since strategic practices are based on good management practice, they should be fairly stable over time. Operational practices are more subject to changes as technology advances and new or updated practices arise to deal with those changes.

## 3.2 Phase 2: Technology View

Phase 2 is focused at the enterprise's technological infrastructure and the questions:

- Do we know enough about our networks to undertake a technology evaluation?
- How do people access each critical asset?
- What infrastructure components are related to each critical asset? What are the key components of the computing infrastructure?
- What technological weaknesses expose your critical assets to threats?
- Which technological weaknesses need to be addressed immediately?

- Strategic Practices
  - Security Awareness and Training
  - Security Strategy
  - Security Management
  - Security Policies and Regulations
  - Collaborative Security Management
  - Contingency Planning/ Disaster Recovery
- Operational Practices
  - Physical Security
    - Physical Security Plans and Procedures
    - Physical Access Control
    - Monitoring and Auditing Physical Security
  - Information Technology Security
    - System and Network Management
    - System Administration Tools
    - Monitoring and Auditing IT Security
    - Authentication and Authorization
    - Vulnerability Management
    - Encryption
    - Security Architecture and Design
  - Staff Security
    - Incident Management
    - General Staff Practices

Figure 6. High Level Structure of the OCTAVE Catalog of Practices

## Do we know enough about our networks to undertake a technology evaluation?

One of the first activities in Phase 2 is to define your vulnerability evaluation strategy. While many enterprises routinely perform their own technology vulnerability evaluations, many more outsource this task and still others only perform this type of evaluation on a sporadic basis. The strategy you chose is largely based on your current capability for performing this type of evaluation and interpreting the results. Have you ever run vulnerability tools? Do you know how to interpret the results? If the answers to these types of questions is no, you may want to limit this activity to a discussion about components and access paths to determine, in general, how vulnerable your assets are. If you've already done or had a vulnerability evaluation done in the recent past, there may not be a need to redo it; check the results for the key components you're interested in and see what vulnerabilities were found. If you contract for this service, you can ask your service provider to look at specific components and work with you to review the data. If you are familiar with these tools and know how to interpret and use the results, then proceed to identify the key components related to critical assets and run the vulnerability tools.

The key is in the interpretation of results. You need to know not only what the vulnerability is but also how it affects the asset. Could that asset be disclosed or destroyed? During an enterprise-wide, asset-based risk evaluation like OCTAVE you can be completely derailed when faced with thousands of vulnerabilities and information you cannot interpret.

**How do people access each critical asset? What infrastructure components are related to each critical asset? What are the key components of the computing infrastructure?**

By looking at your systems and components in relation to the critical information assets, you can determine where to actually run the vulnerability tools. You may find that you need to run tools on the personal computers your employees use at home for remote access. You might also find it more informative to select several components, such as desktop workstations, and look for patterns relative to vulnerability patching. In this case, you are also verifying some of the information about your security practices. While you may believe that you keep all workstations up-to-date with respect to patches, a spot check on random computers may indicate that this is not true. In this case, your vulnerability management and authorization practice areas may need improvement.

**What technological weaknesses expose your critical assets to threats?**

Another aspect to consider with vulnerability tools is their limitations. Vulnerability evaluation tools identify known weaknesses in technology, misconfigurations of "well known" administrative functions, such as file permissions on certain files or accounts with null passwords, and what an attacker can determine about your systems and networks. Within the *Human Actors Using Network Access Tree*, the branches associated with deliberate motive are those most affected by technological vulnerabilities. Most of the other types of threats are not. By looking at all the threats in the profile that are not affected by technological vulnerabilities, you could see that simply managing technology vulnerabilities may not protect a critical asset. You may also need to consider other types of threats. This is, again, another piece of the puzzle you need to solve to understand how an asset is threatened before you spend all of your resources tackling only one source of threat.

**Which technological weaknesses need to be addressed immediately?**

Whenever you run vulnerability tools, you get reports about what is found. Tool providers rank the severity of vulnerabilities in their own or in generally accepted terms. You need to decide what these vulnerabilities mean to you and your critical assets. For example, it may be more advantageous to fix the severe vulnerabilities that directly affect the more critical assets or a specific aspect of a critical asset, such as the exposure of patient records. This is when the ability to interpret tool results is crucial.

## 3.3 Phase 3: Strategy and Plan Development

Phase 3 integrates all of the information gathered in the first two phases to answer these questions:

- What is the potential impact on your enterprise from each threat? (What are your risks?)
- Which are the highest-priority risks to your enterprise?
- What policies and practices does your enterprise need to address?
- What can your enterprise do to recognize, resist, and recover from its highest-priority risks?

**What is the potential impact on your enterprise from each threat? (What are your risks?)**

When something negative occurs, it can have an impact on your company. A risk is an event (a threat scenario), consequences (impacts on the enterprise), and uncertainty (whether the threat scenario will occur). Risks are analyzed to provide additional information to assist decision makers. An enterprise cannot mitigate every risk because of funding, staff, and schedule constraints. Thus, it is necessary to determine relative priorities. In many risk management processes, both impact and probability are evaluated as a means of deciding which risks to deal with first, if at all. For example, a manager may choose to deal with only high-impact, high-probability risks: monitor medium-impact, medium-probability risks; and ignore low-impact, low-probability risks.

During OCTAVE, risks are created by adding impacts to the threat trees (Figure 7). Impacts are described in terms of their effect on such things as the enterprise's finances, productivity, reputation, and legal aspects. These impacts are evaluated based on criteria specific to the enterprise. Good evaluation criteria are contextual measures defined for your enterprise based on current business drivers, objectives, and priorities. For example, $1,000,000 may be a high impact to you, only a medium impact to another. Reputation may take precedence over productivity.

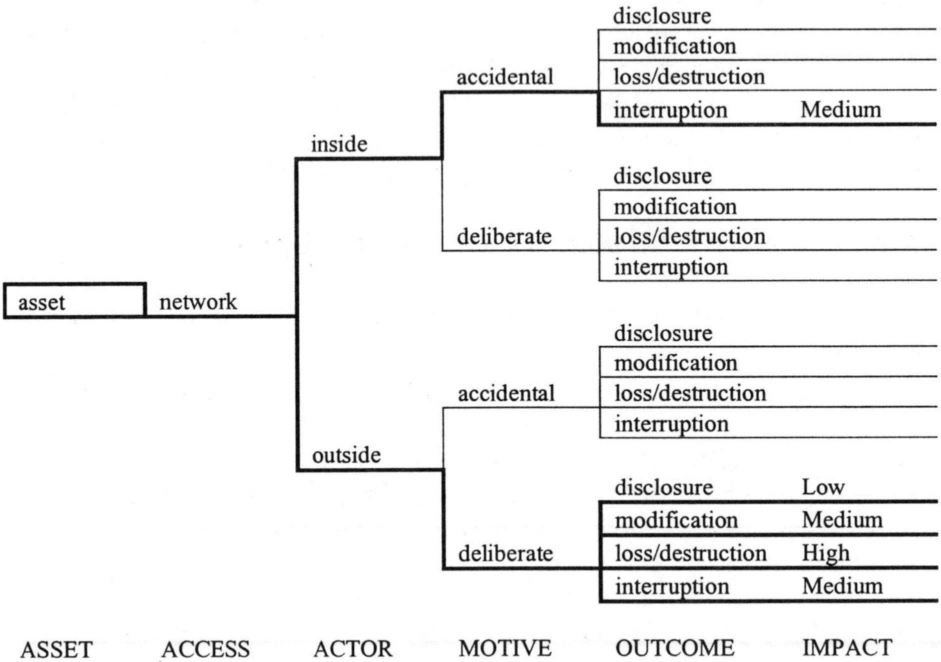


Figure 7. Risks

Impact criteria should also be defined for a set of areas as opposed to one single measure, whether you consolidate all of the values into one or maintain them as separate measures. These areas provide more information on which to base later decisions. A good set of impact evaluation criteria tend to remain stable from one year to the next. The areas of impact in OCTAVE are

- reputation/customer confidence
- life/health of customers
- fines/legal penalties
- financial
- productivity
- other

Qualitative measures of high, medium, and low provide the simplest means of assigning values to risk impacts. Quantitative criteria are hard to define accurately and are hard to use consistently. However, they can be embedded in qualitative criteria, such as defining a *range* of financial impacts. For information security risks, probability is a more complex and imprecise variable than is normally found in other risk management domains. While it is easy to determine the probability of a finite event such as a port scan, determining how that event can be abstracted into a calculation of the probability of someone using the network to attack a specific asset is more difficult.

**Which are the highest-priority risks to your enterprise?**

Priority can be determined by looking at the impact values of the risks (and, if it exists, the probability). The simplest means of establishing priority is defining the highest impact risks as the highest priority. More complicated methods can also be used, such as voting or assigning weights to the different areas of impact. You may decide, for example, that your reputation must remain untarnished while increases in production costs can be accepted. Prioritization of risks is used to help you focus on what needs to be mitigated and improved first. In addition, by improving the security practices associated with critical risks, you also improve and mitigate the less critical risks. A simple gap analysis can point out additional actions that would help mitigate the remaining risks once the critical risks are addressed.

**What policies and practices does your enterprise need to address? What can your enterprise do to recognize, resist, and recover from its highest-priority risks?**

Once the risks have been defined, the process moves toward building the final set of OCTAVE outputs:

- an enterprise-wide protection strategy that defines organizational direction, based on adding or improving security practices
- mitigation plans to reduce the risks to critical assets, using security practices as a means to identify specific mitigation tasks
- short-term actions for immediate, small improvements

A protection strategy pulls it all together to provide guidance and direction for managing security at both the enterprise level *and* the information technology level. The strategy also provides direction for future information security efforts in the enterprise. It defines the activities that an enterprise uses to enable, initiate, implement, and maintain security.

Mitigation plans are created by deciding which risks need to be mitigated, deferred, or accepted. Deferred risks should be mitigated at some point but cannot be dealt with now because of resource constraints, for example. Both the protection strategy and mitigation plans are oriented around the security practices. A protection strategy will focus on organization-wide security practices, such as revising employee hiring/termination policies to include security considerations. Mitigation plans also focus on security practices, but at an asset-specific level, e.g., training a selected number of IT personnel on newer security tools to help monitor a key system.

The practices are chosen based on the high priority risks for a critical asset, however, if you looked at the other risks to that asset, you would see that many of these are also mitigated. The same philosophy holds when looking across at other assets. Focusing on the critical assets provides improvements for other assets.

There are other types of results from OCTAVE, beyond the protection strategy and mitigation plans, that are important to the effective management of information security risks. Workshops produce a strong side effect of team building and increased security awareness. For example, IT personnel realize what users are really doing, users have a better appreciation for security measures, and managers have a better sense of what's really going on in the enterprise. Simply identifying critical assets can change the focus of many other activities and alter resource allocations. The surveys alone produce institutional learning when you understand what you do well and what is missing or only partially performed, and vulnerability evaluations become more useful when they are focused on the parts of the infrastructure that support critical assets.

## 4. WHAT HAPPENS AFTER AN EVALUATION?

As stated in the beginning of this chapter, *evaluating* information security risks is a part of *managing* information security risks. OCTAVE is a slice of the information security risk management cycle (Figure 2). Once the evaluation is complete, an enterprise must still implement the mitigation plans and track them to completion. As with any type of evaluation, the ultimate key to success is to implement the recommendations – not put the evaluation results on the shelf. An effective information security evaluation should support business continuity, enable you to protect your critical information and systems, and provide a foundation for future security improvements.

# 5. CONCLUSIONS

Risk evaluation and management is one way to look at information security. It provides a more proactive basis for dealing with a rapidly changing domain than simply reacting to the latest word of a new vulnerability. While vulnerability patching can never be ignored or pushed aside, it must take place within the broader organizational security practices. We are long past the point at which information security can be safely relegated to just the information technology department. As with many other types of business risk, the enterprise itself must become involved. Finally, we must move beyond the focus on technology, and look instead to what we must protect to ensure the success of the enterprise and how best to use both technology and organizational practice to ensure our survivability.

# REFERENCE LIST

[1] Alberts, Christopher & Dorofee, Audrey. *OCTAVE Criteria, v 2.0* (CMU/SEI-2001-TR-016). Pittsburgh, PA: Software Engineering Institute, Carnegie Mellon University, December 2001.

[2] Alberts, Christopher & Dorofee, Audrey. *Managing Information Security Risks*. Addison-Wesley, Boston, MA. 2002

[3] Alberts, Christopher & Dorofee, Audrey. *Operationally Critical Threat, Asset, and Vulnerability Evaluation Method Implementation Guide, v2.0.* Pittsburgh, PA: Software Engineering Institute, Carnegie Mellon University, June 2001.

[4] Alberts, Christopher & Dorofee, Audrey. *OCTAVE Catalog of Practices, Version 2.0* (CMU/SEI-2001-TR-020). Pittsburgh, PA: Software Engineering Institute, Carnegie Mellon University, October 2001.

[5] Health Insurance Portability and Accountability Act (HIPAA). "Security Standards Final Rule" Federal Register, vol. 68, no. 34, (February 20 2003): pp. 8333-8381.

[6] *Information Security Management: Code of Practice for Information Security Management of Systems* (BS7799). London, England: British Standard Institution, May 1999. (also known as ISO17799).

[7] Howard, John D. & Longstaff, Thomas A. *A Common Language for Computer Security Incidents* (SAND98-8667). Albuquerque, NM: Sandia National Laboratories, 1998.

[8] Hutt, Arthur E.; Bosworth, Seymour; & Hoyt, Douglas B. *Computer Security Handbook*, 3rd ed. New York, NY:John Wiley & Sons, Inc. 1995.

[9] Parker, Donn B. *Fighting Computer Crime*. New York, NY:John Wiley & Sons, Inc. 1998.

**About the Author:**

Audrey Dorofee is a senior member of the technical staff at the Software Engineering Institute (SEI). She is continuing the work to refine and transition OCTAVE to the user community. She is co-author of *Managing Information Security Risks*, the *OCTAVE Method Implementation Guide,* and the *Continuous Risk Management Guidebook.*

Chapter 10

# AUTOMATIC SECURITY POLICY MANAGEMENT IN MODERN NETWORKS

*A New Approach to Security*

S. Raj Rajagopalan
*Telcordia Technologies*

**Abstract**: Problems in network security management have outgrown the capabilities of current human-intensive administrative methodologies. There is a significant need for automation of network security operations that is not being addressed. This article proposes a new way of looking at the security automation problem and lays down some requirements for developing appropriate technologies to minimize human involvement. We also describe a new technology built on this design called "Smart Firewalls."

**Keywords**: network security management closed-loop automation

## 1. MOTIVATION

Security is often cited as one of the biggest cost items in maintenance of enterprise networks today [8]. A cursory look at modern security administration practices shows that our ability to manage network security has not kept pace with advances in networking technology. While technologies for building large-scale networks and network services have advanced dramatically, creating new vulnerabilities and opportunities for complex attacks, systematic principles for network management, especially security management, have lagged behind. As a result, today's state of the practice of security management is highly human intensive and consequently, slow and error-prone. Existing tools have been designed for static security and are inadequate to meet the current demands of user mobility and diversity requiring frequent and error-prone reconfigurations. To quote [1], "Today's market for security products and services is best characterized as a Rube Goldberg mixture of parts, most of which are unable to work with each other." To further worsen matters, there are no tools to verify the correctness or comparability of today's technologies. As a result, it is overwhelmingly true that security policy administration is done on a piece-meal and ad hoc basis by network and security administrators managing firewalls, intrusion detection systems, routers, switches and other network elements in individual sub-networks with minimal co-ordination.

Administrators, balancing the demand for new services with the potential for new security vulnerabilities that can arise, must make decisions based on uncertain and rapidly changing information about the networking environment. Even as security features are added to the newer generations of applications, security breaches continue to plague us by exploiting any avenue possible at the network and lower layers, operating system errors, or back-end databases in ways that effectively thwart these higher-layer security mechanisms, making it critical to have a multi-layered defense. Meanwhile, the volume and complexity of needed management actions have gone beyond human capability to comprehend the mass of network sensor data and respond in a timely and effective manner to every potential intrusion. Frequently, in commercial networks, availability trumps security – risks are taken for fear that patching vulnerabilities may make a critical service unavailable to customers. On the one hand, technology and personnel changes make it nearly impossible to even validate that the current state of a large network upholds any given security policy. Network elements frequently have complicated user interfaces and obscure dependencies between each other such that system administrators have to learn and manage. On the other hand, the introduction of new services such as VPNs, wireless LANs, and SOAP while providing useful new capabilities severely test the capability of existing security mechanisms by breaking old assumptions (such as "inside" and "outside" the network). Add to this the constant churn of new security vulnerabilities in old and new technologies, we simply do not have an accurate picture of the state of security in our networks and are currently unable to address this problem in a reasonable way.

The ad hoc and human-intensive nature of security maintenance drives up the cost of network operations. The position maintained in this paper is that automated self-management will be the defining characteristic of next generation security technology. Our observations have led us to believe that the only solution to today's security problem is to manage networked systems automatically in such a way that desired high-level security goals are upheld in large and dynamic networks with minimal human intervention.

# 2. STATE OF THE ART

Comprehensive security automation is a relatively new area of research. We can split the focus into three related areas of interest and technology: security policy specification, enforcement techniques, and systems integration issues.

## 2.1 Security Policy Specification

It needs no argument that all automation starts with goal specification and in system automation, policy-based management has become a popular topic of research in the last few years (see, for instance, {ref network-issue}) and security policy specification has been treated as a special case of general policy-based management. The most prominent forum for presentation of work-in-progress is the Policy Workshop series (www.policy-workshop.org) and there have been two contenders for the Security Policy Language standard. The first is the Security Policy Specification Language (SPSL) [4], which is an IETF Working Draft. The second is embodied in a sequence of papers by Morris Sloman and his co-workers on the "Ponder" language [5]. Both languages are based on the event-condition-action paradigm of policy. Briefly speaking, this paradigm allows the specification of rules of response in an automated system. Each rule specification states that on the detection of a specified event, the system performs the specified action if the specified condition is true. Either the event or the condition specification is allowed to be null. The genesis of this class of rule-based languages to control systems can be traced independently to database triggers 8 and to discrete system control theory [7]. The advantage of rule-based languages is that they are seemingly easy to specify by direct translation from human patterns of behavior but as we will discuss in a later section, human behavior is not necessarily the appropriate model to emulate in automation. Ironically, the main strength of these languages may also be their drawback in automation, namely, their power of expression. For example, SPSL is powerful enough to allow specification of rules that may cause the enforcement system to go into an infinite loop. The Ponder language allows for a rich set of security rule specification including both mandatory and discretionary policies. Ponder uses a specification environment with a object oriented flavor where rule sets may be defined in a rich variety of relationships between objects controlling access to resources, enforcing audit actions, etc. However, the object-oriented nature of Ponder policies makes it computationally hard to analyze the composite effect of policies on a system. For example, it may be possible to specify policies that are internally inconsistent but this fact may not be uncovered until the policies are actually implemented. However, it is important to note that event-condition-action rules in some form are always to be found in the system because device drivers are typically written in this way because the rules can be very simple.

## 2.2 Enforcement Techniques

Enforcement techniques provide the bridge between policy specification and primitive operations (such as device configurations) that are available in any given networking environment. The main challenge in enforcement techniques is the accurate mapping of policies to primitives and depends critically on the difference in the semantics between policies and primitives. Obviously, when the two are very close mapping is easy but the power of automation is limited and vice versa. On the one extreme, SPSL is a very expressive language and computing responses to events or conditions may take a very long time to compute. The Ponder policy language on the other hand directly maps policy language constructs into network primitives with little or no computational cost because the difference between the two descriptions is small. In most practical environments such as enterprise networks, the need for automation is somewhere in the middle. However, this spectrum is not well explored. This is not to say that no work exists. There is a very large and rich tradition of verification of security policies for software engineering (see, for example, [6,11]), which unfortunately is not easily translated into mechanisms for policy maintenance for network systems. The most significant advance in this regard was "Firmato" [2], which maps a sophisticated role-based security model into firewall rule-sets. Firmato limits itself to firewalls and it does not seem to extend to other devices or settings where firewalls are absent (such as in the interior of enterprise networks).

## 2.3 Instrumentation and Systems Integration:

The power of policy enforcement ultimately lies in the actual primitive monitoring and control mechanisms that are exercised by the system. Automation merely provides a layer of management atop these primitives and allows users to effectively leverage their power to achieve system goals. When it comes to system-wide automation and control, network instrumentation is very fragmented. The only generic and widely supported platform for network monitoring (and some control) is SNMP [11], which is a very simple protocol not designed as an abstraction for cross-network automation. [1] provides an incomplete view of what is needed in systems integration and automation but it is a step in the right direction. Basic network instrumentation falls into three categories:

(a) Monitoring: Programs like HP Open View and What's Up that monitor various devices provide collected status information on networked systems.

(b) Configuration: Programs from Cisco, Ecora, and Solsoft etc that help push configurations to various devices like firewalls, routers, switches etc.

(c) Audit: Programs such as traffic analysis tools and Intrusion Detection systems that report anomalies in traffic.

However, it has to be noted here that the technologies in these three categories are rarely integrated.

## 3. HOW DO WE GET AUTOMATED SECURITY

Network security management tools need to automate management of network security in dynamic environments to the fullest extent possible. Rather than depend on human administrators to provide the right configurations on each element in any situation, it would be necessary to enable network elements such as firewalls, routers, and switches to adapt to change by reconfiguring as appropriate. The challenge would then be to compute the right reconfiguration for the network as a whole so that the appropriate security policies are upheld without causing loss of service to legitimate users. Note that because of the increased interdependence of network elements and services, deployment of a new service for a user and conversely, the denial of that service, can involve reconfiguring multiple network elements and hence it is all the more important to consider the configuration of the entire network rather than just individual devices.

Considering the reality that security is no longer just a matter of protecting a corporate network from the Internet but also protecting one sub-network from another, these reconfigurations have to include routers, switches, workstations, servers, etc. Furthermore, one of the specific goals of this work is management of security configurations in networks that span multiple administrative domains. This has become a real need in today's world of temporary coalition between (otherwise adversarial) corporations. This is also true within networks of large multi-national organizations that are often administered by different people with different views of privileges and responsibilities. A paradigmatic example is the situation of two connected firewalls, each of which has a local security policy (administered by one or more administrators). Even if each firewall correctly implements the local policy, the interconnection of the two firewalls may violate a global security policy that neither firewall can detect by itself. Clearly, we need to be able to reason about a large network to verify whether the totality of the local configurations upholds or violates global security goals. Simply put, our goal is to be able to answer questions such as "Can these two sub-networks be connected to each other without violating either security policy?" Such questions cannot be answered today with any reasonable degree of certainty. Indeed, in most cases, even the principles that should guide such important decisions are not clear. In the following, we attempt to delineate what some of these principles should be.

## 3.1 Principles for Security Automation

Here we present a set of requirements that any technology for security automation needs to satisfy in order to be acceptable to network operators and executives alike.

1. High-level goal specification: For automation to be effective, the user (administrator or network owner) must be able to specify the network security intent easily and effectively. Typical security goals tend to be at a very high level compared to what a language like Ponder would support. Except for specific cases like CORBA servers, system-wide security goals do not fit easily into the object model. Security goals typically specify access restrictions to network services for user categories. For example, a security policy may specify that guest users may only access email in the DMZ. Another may specify that employees should always have access to the timesheet server. Policy should be as close as possible to the statement of intent but also formally precise enough that they can be input to a program. For this reason, policy languages that allow general statements in English as policies are not useful.
2. Policy Auditability: True security automation implies the ability to audit for policy compliance, i.e. check whether all policies are being upheld by the system at any moment. ECA rules do not lend themselves easily to such audit since they are, by definition, triggered by events. However, audit may be possible, if onerous, by maintaining the history of relevant events. Ideally, the concept of policy should support a snapshot view of policy compliance.
3. Policy Conflicts and Enforceability: In a realistic environment, policies will be generated by multiple sources and internal conflicts between policies are possible. Depending on the power of the language, conflict resolution may be a computationally

hard problem and policy language may have to be restricted to make conflict detection (and resolution) feasible.

4. Policy Validation: Given any network description and a set of policies, it must be possible for a program to automatically and quickly verify that the policies are valid in that network. This is why the computational cost of validation must be taken into consideration in language design. As mentioned earlier, the proposed standard SPSL has an unacceptably high cost of validation due to its complexity. In general, almost all logic-based languages that allow unbounded quantifiers suffer from this problem. Policy language thus has to be as simple as possible and support most if not all the needs of network security specification.
5. Synthesis: Discovering when a network is in violation of policy is necessary but not sufficient for automation. It must also be possible to synthesize changes to the network that will bring it back to compliance safely and quickly. This is trickier than it sounds. It is possible, even likely, that there are many possible options exist to bring a network into compliance. Administrators use a variety of homegrown heuristics to rank these options and it may be hard to capture these preferences in a program.
6. Closed loop automation: Finally, security policy automation will only be achieved when an automation tool detects changes, validates the changes for policy violations, synthesizes new changes to force compliance, implements these changes on the network, and listens to the network for new change to close the loop. The challenges in this space are mostly concerned with control issues such as stability. While these problems are hard in general, the author believes that useful applications can still be designed by intelligent tradeoffs of complexity and automatability. Section 4 describes a project that this author led and the design decisions that address these tradeoff issues.

## 3.2 Some Examples of Policy Automation

To give the reader a feel for the kind of automation we feel is desirable, here are some example situations where automation would help an administrator, Chief Security Officer, CIO, or whoever has operational responsibility for network security. It would be desirable for software to answer questions like the following:

- Is it safe to provide "mail" service to "guests" on the "internal" network? If not, what security policy would be violated by this action and how?
- We want to bring the "Web Server" "W" in sub-network X down for maintenance without users losing service. Find an alternate mirror web server in sub-network Y that does not violate any security policy and reconfigure routers and firewalls appropriately. Note that this form of policy allows us to alter in a modular fashion the meaning of "Web Server" itself from today's Apache, IIS, etc. to new technologies when the need arises.
- A guest laptop has connected to port n of switch S. Is this policy-compliant under the current configuration? If not, compute a correct reconfiguration of the switch and other network elements as necessary so that this laptop has access to all the services that are allowed for guests but does not see any internal machines it is not supposed to.
- A new security-alert says that "ftp" servers of version x.y can be hacked into and provide root access to the hacker. Reconfigure the network so that this vulnerability cannot be used to violate any security policies.
- We want to add a new kind of device to the network with the specified connectivity. Validate security policies in the new network and report any reconfigurations for this switch and other elements that would need to be done to incorporate it safely.
- The network is connected by a specified connection to another network belonging to company C that has its own security policies. Reconfigure so that our network cannot be used to violate company C's policies.

## 4. THE SMART FIREWALLS PROJECT

We describe here the work done in our project titled Smart Firewalls (see [3] for more technical details), funded by DARPA under the Dynamic Coalitions program to design, develop, and demonstrate a system for automatically managing security policies in dynamic networks. Specifically, we aim to reduce human involvement in network management by building a practical network monitoring/reconfiguration system so that simple security policies stated as positive and negative invariants are upheld as the network changes. The focus of this project is a practical tool to help systems administrators verifiably enforce simple multi-layer network security policies. Our key design considerations are computational cost of policy validation in a dynamic environment and the power of the enforcement primitives. The central component is a policy engine populated by declarative models of network elements and services that validates policies and computes new configuration settings for network elements when they are violated. We instantiate our policy enforcement tool using a monitoring and instrumentation layer that reports network changes as they occur and implements configuration changes computed by the policy engine. While efforts to develop ever more sophisticated firewalls, to ensure the protection of individual end-hosts, and the privacy and integrity of end-to-end communications continue, what is lacking is the ability to address as a whole, network-wide security policy considerations. We propose to automate the enforcement of network-wide security policy at all levels by making security configuration management dynamic and responsive.

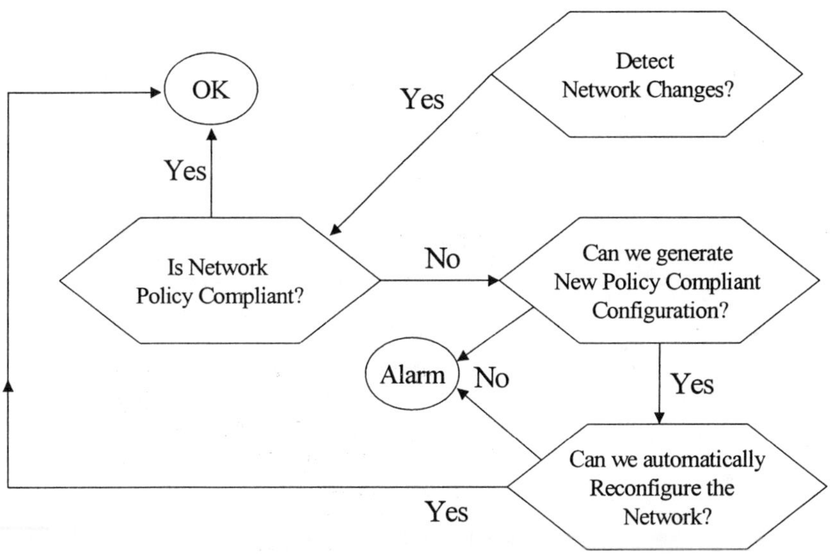


Figure 1. Smart Firewalls Schematic for Policy Management

Our approach starts with a paradigm shift in security policy from expressing policy questions as unstructured combinations of access policies and ad hoc prescriptions for networks elements, to expressing network security policies purely in terms of access goal invariants (both positive and negative) to applications and network services. Given a network, we want to verify that the desired access is enabled and the undesired access is verifiably denied. By building a computationally efficient framework for checking whether security policy is being upheld, avoiding the state explosion problem with traditional formal verification approaches, we make the automation of policy administration feasible. Smart Firewalls has been validated in multiple experiments including the US Army Future Combat Systems capability demonstration and the Joint Warrior Interoperability Demonstration. Smart Firewalls is undergoing a process of technology transition to commercial enterprise networks.

We start the description of our prototype by listing the fundamental questions that are listed in the flow diagram in Figure 1. To answer these questions in an automated framework, the Smart Firewalls uses three decoupled components that are described in brief below (see Figure 2). Figure 3 shows the flow of information between the components.

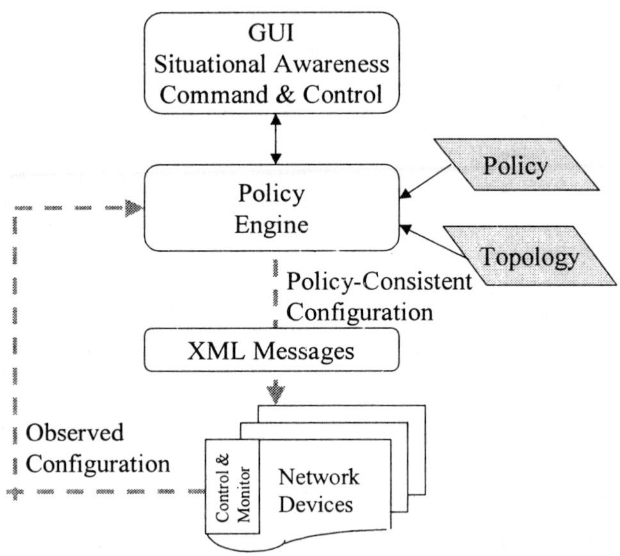


Figure 2. Smart Firewalls Architecture

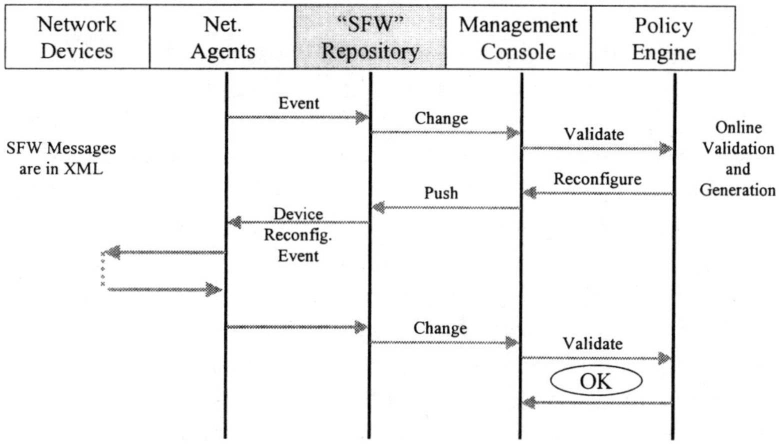


Figure 3. Information Flow in Closed-loop Policy Management

In summary, the chief novelties of this project are:

(1) Policies representing end-to-end security requirements as who should have access to what service.

(2) Validation algorithm that analyzes an arbitrary network holistically (rather on a per-device basis) for policy violations.

(3) Closed loop mechanism that synthesizes and pushes reconfigurations automatically when policy violations are detected.

(4) Support for local autonomy so that administrators are allowed to make local changes to configurations as long as global policies are upheld.

(5) Planning and change control tool that performs impact analysis of proposed changes on policies.

## 4.1 Policy Language

Our Policies are expressed independently of the underlying topology as invariants that should be true irrespective of network changes. At present our policies are four tuples <Client_tag, Server_tag, Service_tag,

Allow/Deny> where the tags are logical names that can be dynamically associated with physical entities in a particular network. Some examples of security policies: (1) <Non-Employee, Internal_Server, Web_service, Deny> (2) <Guest, DMZ_Network, Mail, Allow>. In addition to these policies, we also allow meta-policies such as (1) Default Deny (Every service that is not explicitly permitted is prohibited) and (2) Local policies are upheld if they are not in conflict with global policies. Once such policies and meta-policies are specified the job of our prototype is to ensure that the current configuration of the network always makes these policies true. Note that these policies do not specify how they are to be upheld. That decision is made on the actual network presented and the constraints therein. Also note that we allow logical tags in our policies such as "DMZ_network" in the server tag above. The semantics of "Allow" policies state that an actual server satisfying the tag and providing the requisite service has to be found in the given network and the network configuration managed accordingly. Similarly, the semantics of "Deny" policies state that all actual entities satisfying the client tag are denied access to the specified service on the specified server tag. A service tag such as "Mail" may be mapped to IMAP in one context and POP in another without changing the policy statement.

## 4.2 Network Instrumentation

Policy management automation depends crucially on being able to sense and control the given network automatically. For the purposes of the initial demo we used the NESTOR system [13] from Columbia University. The current design is simpler and relies on external network management platforms such as HP Open View and adapters for commercial firewalls. The adapters support a uniform abstraction for network devices such as routers, switches, and firewalls, as well as end-hosts such as clients and servers, providing both monitoring and reconfiguration functionalities for these devices that use the same abstraction. Adapters are merely translators between native control data formats and XML. Adapters post discovered changes and receive reconfigurations by exchanging XML messages with an asynchronous bus. An important note to make here is that we do not assume that a reconfiguration is successful as pushed. A reconfiguration may fail or may be altered for a number of reasons. We simply monitor the network again after a suitable amount of time and revalidate the results.

## 4.3 Policy Engine

The Policy Engine is the core of Smart Firewalls technology. It logically validates a given network configuration against the specified policies using internal models of network devices and computes whether each policy statement is true in the network as a whole. Whenever the network as a whole is deduced to be in violation of one or more policies, the Policy Engine also attempts to compute a reconfiguration that will make the network policy compliant. In addition to policy management, the Policy Engine can also answer queries about the availability of services in the network. The policy engine needs to handle the large space of possible states, and respond to network events quickly and correctly. Using declarative reasoning the policy engine avoids exhaustive enumeration of good states. Avoiding the use of model checking which tends to be computationally expensive, policy invariants are mapped into connectivity properties of an abstract, edge-colored graph which represents the totality of the given network topology and configuration. By efficiently computing transitive closures on this abstract graph, the policy engine is able to validate policies on large networks rapidly without pre-processing. This enables us to perform policy validation on demand and respond in near real time for moderate-sized networks. Declarative models of network elements enable us to capture not only normal functions but also flaws in devices and software. Using models of flawed entities allows us to detect security breaches caused by discovery of previously unknown defects as in CERT alerts (www.cert.org) or by the introduction of new network elements. The policy engine employs a two-step process of configuration generation followed by validation. The configuration generation phase takes a partially configured network and fills in any missing parts of the configuration. The combination of the completed configuration and the network topology is then validated against the policy. Using model composition, the policy engine can handle some feature interaction problems caused by interconnected networks of devices.

## 4.4 Management Console

The management console is the user interface to the policy engine and the network and allows policy inputs, network descriptions and visualization, and query/response. The console is also used for auxiliary inputs such as security alert information and models of new networking devices. The functions of the console are to: (1) report on the current state of the network being administered in a human–comprehensible form, display network changes that have been detected, and displays warnings as necessary. (2) Accept control information from systems administrators in the form of policies, models, or networking constraints. (3) Store persistent information in XML to eliminate the need for repeating the same input. (4) Allow the user to pose queries on the network state.

# 5. CONCLUSIONS AND FUTURE WORK

Automating security administration is the key security challenge today. Academia and industry are only beginning to address some of the issues that are central to this challenge. The issues that need to be addressed include: systematic methods for evaluating and monitoring security properties of large-scale networks, tools for managing configurations of network elements such as firewalls, switches, routers, application servers, etc. in a large network, and the capability for networks to self-reconfigure automatically in response to changes in the network while maintaining global security properties. We describe a project called Smart Firewalls that addresses some of these questions and provides a capability beyond current practice. This article has not addressed related areas of interest such as user and application security management that are much better studied but focused on automation of *network security.* We feel that this area will be worth watching for breakthroughs both in research as well as commercial products.

# ACKNOWLEDGEMENTS

This material is based upon work supported by the Air Force Research Laboratory under Contract F30602-99-C-0182. The views and conclusions contained in this document are those of the author and should not be interpreted as representing the official policies, either expressed or implied, of the Air Force Research Laboratory or the U.S. Government.

# REFERENCES

1. Aberdeen Group. Security Process Automation, An Executive White Paper. May 2002. http://www.aberdeen.com/2001/research/06020003.asp.
2. Y. Bartal et al. Firmato: A novel firewall management toolkit. Proc. IEEE Computer Society Symposium on Security and Privacy. 1999.
3. J. Burns, P. Gurung, D. Martin, S. Rajagopalan, P. Rao, D. Rosenbluth, A.V. Surendran. Automatic Management of Network Security Policy. Proc. DISCEX 2001. Also at http://govt.argreenhouse.com/smartfirewalls.
4. M. Condell, C. Lynn, and J. Zao. Security Policy Specification Language (SPSL). Internet draft, Internet Engineering Task Force, July 1999.
5. N. Damianou, N. Dulay, E. Lupu, and M. Sloman. The Ponder Policy Specification Language. Policy 2001. Pp 18-39.

6. L. Gong et al. Going Beyond the Sandbox: An Overview of the New Security Architecture in the {Java Development Kit 1.2}. USENIX Symposium on Internet Technologies and Systems. Pp 103—112. 1997.
7. R. Huuck et al. Combining a Computer Science and Control Theory Approach to the Verification of Hybrid Systems.
8. L. Koetzle, C. Mines, M. Porth, and H. Liddel. IT Security's Awkward Adolescence. Forrester Research TechStrategy Report. August 2002.
9. A. Silberschatz, H. F. Korth, and S. Sudarshan. Database Systems Concepts. McGraw-Hill. 2001.
10. R. Sreenivas and B. Krogh. On Condition/Event Systems with Discrete State Realizations. Discrete Event Dynamic Systems: Theory and Applications 1, pp 209-236. Kluwer Academic. 1991.
11. T.Y.C. Woo et al. SNP: An interface for secure network programming. Proceedings of USENIX Summer Technical Conference. 1994.
12. Z. Xu, B. P. Miller, and T. W. Reps. Safety Checking of Machine Code. SIGPLAN Conference on Programming Language Design and Implementation. 2000.
13. Y. Yemini, A.V. Konstantinou, and D.Florissi. Nestor: An architecture for self-management and organization. Tech. Rep. Dept. of Computer Science, Columbia University, Sept. 1999. http://www.cs.columbia.edu/dcc/nestor/.

**About the Author:**

Dr. S. Raj Rajagopalan is a Senior Scientist in Security Research at Telcordia Technologies, Inc. in Morristown, New Jersey. His research interests include Cryptography and Security. Since 1999, he has been working on a security automation project with DARPA funding called Smart Firewalls that he is currently helping commercialize. In 2002, he received the Experimentation Pioneer Award from DARPA and the Telcordia CEO Award for his work on security policy management.

Chapter 11

# THE INTERNAL REVENUE SERVICE'S MISSION ASSURANCE PROGRAM

Len Baptiste
*Retired, Former Chief, Security Services, IRS*

**Abstract:** The Internal Revenue Service (IRS), a branch of the Department of the Treasury, is a large Federal agency with over 100,000 employees and 750 facilities. In 2000, it collected more than $2 trillion in revenue and processed 226 million tax returns. In carrying out its tax administration responsibilities, the IRS is committed to adequately protecting the confidentiality of the taxpayer records that it processes. Taxpayer privacy rights are ensured through the Privacy Act, other federal statutes, the Internal Revenue Code, and other IRS policies and practices. The IRS' security approach focuses on a mission assurance program that assures the confidentiality of taxpayer records; the safety and health of its employees; and the protection of its tax administration operations, including its systems, facilities and other resources.

**Key words:** IRS, Security Organization

## 1. INTRODUCTION: THE IRS INITIATES ACTIONS TO IMPROVE SECURITY

In an effort to improve its security, the IRS initiated an internal task group to assess the effectiveness of its security management capabilities in the mid 1990s. The task group was to address findings by its internal audit office and the U.S. General Accounting Office (GAO), which continued to identify security deficiencies. They found that there was no comprehensive program in place to ensure the adequacy, consistency and effectiveness of security throughout the Service. One of the group's recommendations was for the IRS to consider establishing a centralized leadership position to adequately address the security program shortcomings. During a subsequent GAO review [1], which resulted in recommendations to the IRS for strengthening the effectiveness of its computer security management, the IRS established an executive-led security organization to establish and enforce standards and policies for all major security programs including, but not limited to, physical security, data security, systems security and personnel security.

Currently called *Security Services*, this organization was not established to duplicate audit and evaluation reviews performed by the General Accounting Office (GAO) or the Treasury Inspector General for Tax Administration (TIGTA). Instead, Security Services uses some of the same evaluation disciplines employed by auditors to evaluate and baseline current security conditions at various IRS facility types. Security Services works with the IRS business process owners and support functions (e.g., Information Systems and Real Estate) to develop and implement corrective action plans, which it then oversees to ensure progress. These plans focus on developing sound, workable and steadfast security processes. To ensure the adequacy of the security leadership and guidance provided by Security Services, this organization is staffed with a team of experienced senior security and evaluation managers who have the skill mix that is needed to lead and strengthen security across the IRS. Contractor support is also utilized in areas where even more specialized skills are needed to help the IRS to institutionalize critical enabling capabilities, such as its computer security incident response capability (CSIRC).

In 1997, Security Services began focusing its efforts at its 12 large centers. After a three- to six-month period to initially staff the organization, work progressed in a timely manner with the GAO reporting in 1998 that the IRS had adequately mitigated about 63 percent of the weaknesses reported in 1997. In this regard, a senior GAO official noted that this effort was unprecedented and that one third as many corrective actions would have been considered a significant accomplishment. By 2000, over 90 percent of these weaknesses were adequately addressed. The GAO later reported to the new Congress and Administration that computer security was a continuing high-risk area throughout the federal government, but it noted that the IRS was an agency that had taken significant steps to improve its security capabilities by correcting a significant number of weaknesses and establishing an agency-wide computer security management program that should, when fully implemented, help it to effectively manage its security risks [2]. In subsequent years, work branched out to hundreds of other IRS facilities, and continues to address the needed corrective actions.

Concurrently, another important initiative was taking place in 1997 that was directed at deterring, preventing and detecting unauthorized access to taxpayer data and ensuring that consistent disciplinary actions are taken when unauthorized accesses are proven. This initiative resulted in a number of actions, including a comprehensive awareness program that provides annual agency-wide briefings to all IRS employees on unauthorized access, and includes a form that is signed by each employee and their manager to document attendance at a briefing. It also resulted in implementing an improved automated audit trail analysis tool to improve the detection of unauthorized accesses.

The objective in the remainder of this paper is to explain how the Security Service was organized, outline the security practices it developed for use in the IRS, and the factors that contributed to its success.

## 2. ORGANIZATION: SECURITY SERVICES PLACEMENT IN THE IRS

Through April 2003, the Chief of Security Services reported directly to the Deputy Commissioner for Modernization/Chief Information Officer, who headed up the Service's Modernization, Information Technology & Security Services organization, which reported directly to the Office of the Commissioner Internal Revenue. The three executive-led organizations that reported to Security Services were (1) Security Policy Support & Oversight, (2) Mission Assurance and (3) Modernization Security.

*Security Policy Support & Oversight* focuses on working with organizations throughout the IRS to develop, adopt and implement security policies to support operations and protect resources and to perform compliance reviews using the Treasury Security Assessment Framework that is also used to plan and measure security improvements in 15 security capability areas. Particular emphasis has been placed on material weaknesses, lockbox operations, receipt and control operations, and mail handling at campuses.

The *Mission Assurance* organization focuses on enhancing computer security incident response capabilities, coordinating and facilitating situation awareness management center operations, promoting security awareness, coordinating business continuity, and emergency preparedness planning and capabilities, and conducting the sensitive systems security certification and accreditation process. The office also works closely with its peer Security Services offices in monitoring or implementing security policy in existing or modernization systems.

The *Modernization Security* organization focuses on ensuring the adequacy of the security architecture, security engineering, the IRS' Security Technology Infrastructure Release, enterprise life cycle deliverables and transition to support activities. Modernization Security also participates on the modernization executive steering committee and sub-executive steering committees. It also ensures that the modernized security capabilities adequately safeguard taxpayer and other sensitive data.

To ensure that business process and systems owners have a leadership role in deciding on the adequacy of security and associated risks, *a security governance structure* was established at the IRS. Prior to the establishment of this governance structure, concerns were raised that Security Services was driving security solutions that were not always business friendly. The governance structure consists of three executive-level committees—i.e., the *Security Executive Steering Committee*, the *Operations Security Committee* & the *Technology Security Committee.* They are focused on assuring that adequate and consistent policies and guidance are established to safeguard business operations. Their responsibilities include:

• Building and maintaining a coordinated view of IRS-wide security policy and implementation needs, issues & initiatives;

• Enabling effective interchanges between IRS business and security organizations on resource decisions and field impacts associated with security matters; and

• Deciding on security policies with mission assurance as the key driver.

Figure 1 depicts how this governance process is assisting the ongoing initiative to develop and implement consistent security procedures for all operating and functional organizations. In this regard, new directives and recommendations from oversight organizations are channelled down to Security Services and/or the governance committees for action. Security Service plays a lead role in recommending actions with the committees and their working groups, which are established to adequately address new directives and recommendations. Once agreements are reached and commitments made, Security Services is responsible for reviewing the adequacy of the corrective actions being implemented. Subsequently, Security Services is responsible for periodic compliance reviews.

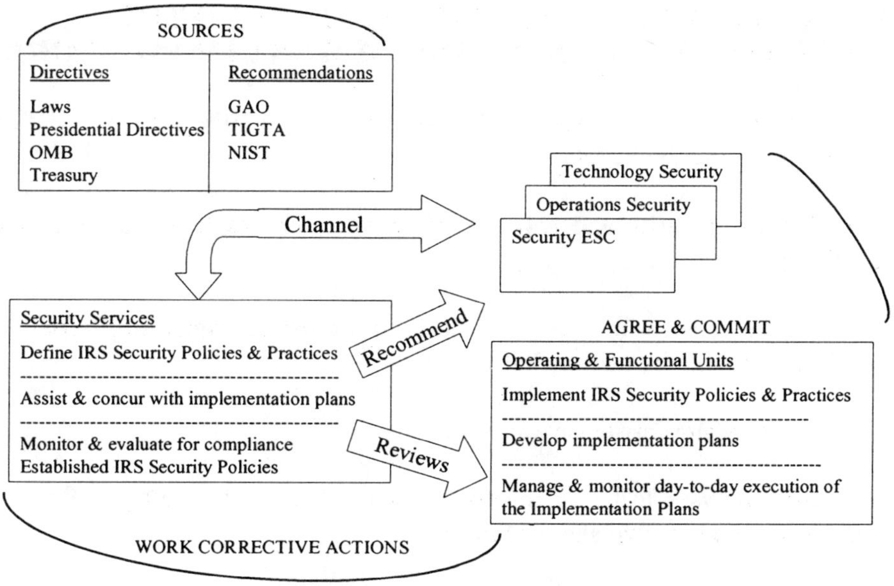


Figure 1. Process used to enhance security procedures.

This entire initiative has grown and matured over the last six years, with many benefits resulting from enhanced capabilities that continue to get even better.

A key to the success of Security Services was the willingness of many IRS managers and staff at the centers and support functions to work closely with Security Services staff to implement the needed security improvements. Because these managers and staff are the key players in institutionalizing the security improvements that are needed, security-training efforts were initiated to enhance the day-to-day security skills needed at the centers and support functions.

This approach has worked well because of the high importance that has been placed on security and privacy by IRS management. Substantial resources have been committed to Security Services since 1997 to adequately establish strong security capabilities that have strengthened the IRS' business processes.

# 3. SECURITY PRACTICES IN THE IRS AND THE TREASURY SECURITY ASSESSMENT FRAMEWORK

Work at the IRS centers and other offices initially focused on key security areas that included:

• Physical security and access mechanisms such as locks, surveillance equipment, and fences;

• Logical security mechanisms that limit access to computing resources to only those with the need to know;

• Data communications management;

• Risk analyses to better identify security threats, their magnitude, and areas needing additional safeguards;

• Quality assurance for reviewing software products and activities to ensure compliance with standards and procedures;

• Security awareness, and

• Contingency planning to ensure the availability of resources and facilitate the continuity of operations in an emergency situation.

As the workload grew to adequately address the various areas needing improvements, so did the Security Services organization. It also developed a security assessment framework, which was modelled after the Federal CIO Council's framework. The Department of the Treasury later adopted this framework for use by all its bureaus. Referred to as the Treasury Assessment Framework, it consists of 15 areas that the IRS uses to drive and track planned and ongoing initiatives and to ensure compliance with requirements driven by laws, regulations and other federal policies and guidance.

The IRS uses the Treasury Security Assessment Framework to drive and measure the effectiveness of its program in the following 15 security areas.

1. Security Policy and Planning: Policies, procedures, roles and responsibilities that characterize and define the agency's security program.
2. Risk Management: An integrated, continuing process to assess and manage security risks in all IRS business processes.

3. Review of Security Controls: Routine review of security controls to assess their effectiveness.

4. Rules of Behaviour: Security rules, roles, and expected behaviours are defined, disseminated & followed.

5. Life Cycle Management: Security is managed through the Systems Life Cycle.

6. Processing Authorizations: Required formal authorizations for system and risk management, including systems security certification and accreditation.

7. Personnel: Effective security-related personnel policies & practices.

8. Physical and Environmental: Appropriate physical & environmental security policies & practices.

9. Computer Support and Operations: Security procedures for computer support and operations are defined and executed day-to-day.

10. Contingency Planning: Contingency plans are established, tested & updated.

11. Documentation: Responsible officials adequately document system-specific security controls & operational procedures for use.

12. Training: Effective security training and awareness is in place to implement the security program plan.

13. Incident Response: A well-defined, timely & effective 24/7 incident response capability is maintained.

14. Access Control: Access to bureau resources are controlled & tested.

15. Audit Trails: Sufficient audit trail policies, procedures & practices are in place to support security objectives to effectively detect misconduct.

In developing the framework, four key components were built into it to provide IRS management with a better understanding of its Service-wide security capabilities.

- Assertions: The statements of the goals and objectives.

- Performance Criteria: The sought-after condition to recognize attainment of the goals and objectives.

- Corresponding Metrics: The defined measures.

- Assessment & Rationale: The current condition, based on vulnerability assessment results.

By using these components, each of the 15 areas is assessed and graded periodically using three color-coded maturity levels: red represents an inadequate capability, yellow represents a developing capability and green represents an adequate capability. This approach has resulted in meaningful progress and has helped to prioritize efforts competing for available resources.

To better depict its ongoing initiatives, Security Services put together what it referred to as its sandwich chart (see Figure 2). Policy and guidance work is performed in the top slice of bread and compliance reviews—which monitor, evaluate and recommend improvements—are shown in the bottom slice. Inside the sandwich, Security Services identified important program operations (e.g., Systems Certification & Accreditation) and value-adding services (e.g., Systems & Telecom Security Standards), which impact all IRS operating organizations. These are depicted in the columns of the chart (e.g., NO refers to *National Office* and ITS refers to *Information Technology Services*).

## 4. THE IRS FOCUSES ON IMPORTANT ENHANCEMENTS AND INITIATIVES

The IRS was able to quickly assess its situation following the tragedies of September 2001. However, this assessment identified areas where enhancements were needed. The following are examples of these enhancements:

- Mail handling locations and operations were improved to better protect employees.

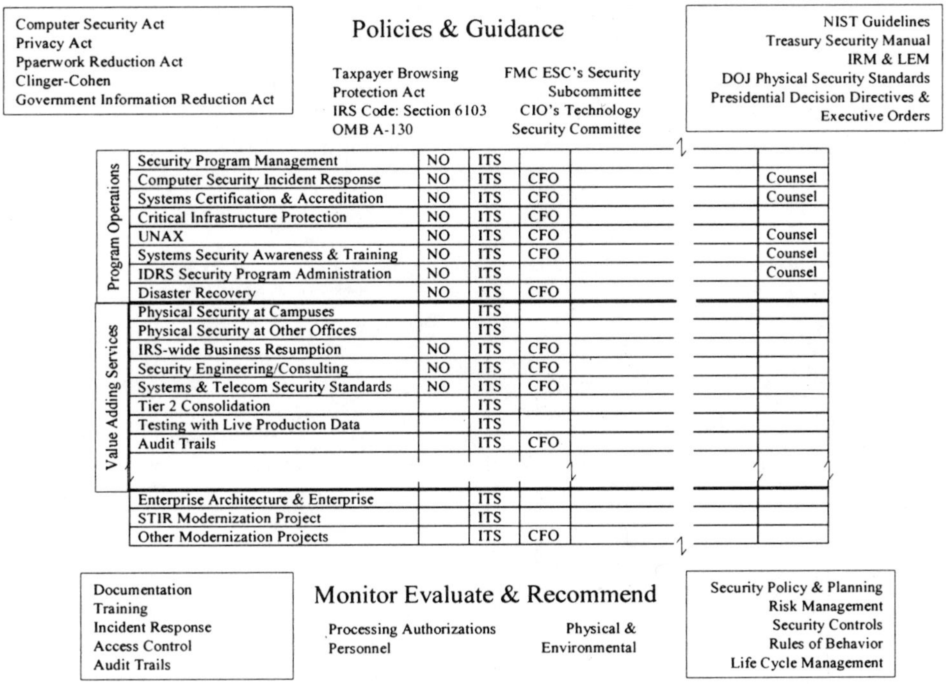


Figure 2. Overall mission and assurance program.

• Decision support capabilities were enhanced with the establishment of situation awareness and management centers.

• The headquarters continuity of operations plan was updated and enhanced.

• Incident response teams were established to quickly deploy needed resources to impacted locations.

• Disaster recovery capabilities were upgraded to more quickly recover critical systems processing.

Other critical security-related initiatives, undertaken by the IRS include actions to mitigate computer security weakness. The current emphasis is focused on (1) developing and implementing consistent security procedures for all operating and functional organizations and (2) ensuring the day-to-day execution of these procedures. Figure 3 depicts the security capabilities life cycle used by the IRS to continually improve its security posture.

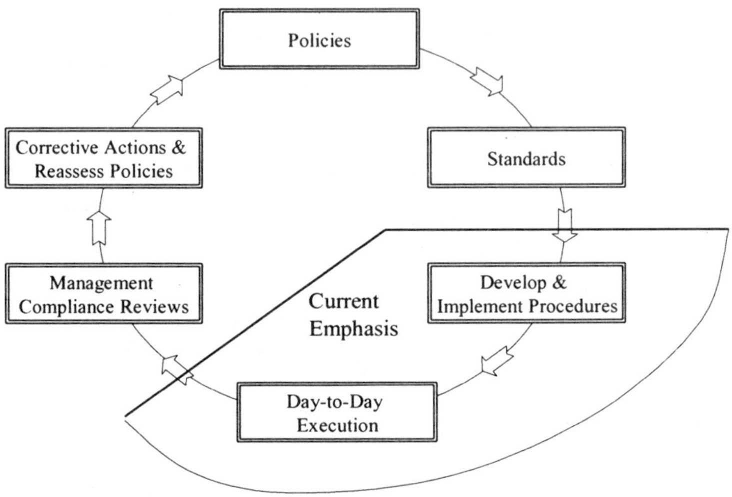


Figure 3. IRS securities capabilities life cycle.

Initial work in 1997 focused on performing compliance reviews to baseline the existing condition, which drove corrective actions and policy revisions. As policies and standards started to get implemented, it became apparent that all stakeholders needed to participate in these decisions to ensure that security solutions were balanced with operational effectiveness. In this regard, a governance structure was developed to oversee security policies and standards.

## CONCLUSIONS

In closing, it is important to understand the following key success factors that were instrumental in establishing a meaningful mission assurance program at the IRS.

- Obtain senior management support;

• Establish a strong team of security and evaluation professionals;

• Identify existing conditions and weaknesses, so they can be adequately managed and monitored;

• Develop corrective action plans with business owners that include both short- and long-term actions, which are prioritized based on risk and available resources;

• Ensure all security disciplines (e.g., physical, personnel and systems security) are addressed in the corrective action plans and if possible develop a security framework to track and measure success; and

• Take advantage of recommendations and lessons learned by others.

Rather than arbitrarily implementing various security capabilities to address risks, the centralized security organization has enabled the IRS to better manage and prioritize the implementation of its security capabilities. In this regard, it has used its various measurement tools (e.g., the Treasury Security Assessment Framework) to track current security conditions and progress. The IRS uses its strong staff of security specialists to perform reviews and help business process owners in developing and establishing corrective actions when needed. It is believed that this organizational approach, and the kinds of assessment tools developed by the IRS, can be adapted for use in many other organizations.

## REFERENCES

1. GAO. IRS Systems Security: Tax Processing Operations and Data Still at Risk Due to Serious Weaknesses, GAO/AIMD-97-49, April 8, 1997.
2. GAO. High Risk Series: An Update GAO-01-263, January 2001.

### About the Author:

Len Baptiste retired in May 2003 from the IRS, where he was the senior executive responsible for all security operations. Prior to joining the IRS, he led systems and security reviews at the U.S. General Accounting Office. He has a BS in Accounting and a MA in Business Management.

Chapter 12

# ORGANIZING FOR SECURITY

Paul Rohmeyer
*Partner & Chief Operating Officer, ICONS Inc.*

**Abstract:** Previous chapters in this book have examined a wide range of technical, legal, and organizational issues with regard to security. This final chapter provides in-depth coverage of the issues involved in developing a secure organization. It starts by examining the forces that are making security a top organizational concern. It goes on to a description of the information security organization from the perspectives of organizational theory and behavior theory. Next, the chapter provides an in depth discussion of the relationship between the CIO and CSO and concludes with a summary and brief discussion of the issues faced by organizations as they develop an architecture for security.

**Key words:** Security, organizational architecture

## 1. INTRODUCTION

The importance of information security to organizations, and society in general is increasing. The growing attention may be attributed to several important trends. The continued adoption of electronic commerce, particularly business to business arrangements is growing at a brisk pace despite a general economic downturn. A large number of organizations recognize electronic commerce as a permanent organizational reality and, therefore, seek to optimize the management of supporting information and technology resources.

The United States National Security Agency (NSA) and National Infrastructure Protection Center (NIPC) have provided direction to organizations that may be considered parts of the nation's critical infrastructure. Presidential Decision Directive (PDD) 63 of the Clinton administration provided recommendations concerning information security to the leaders of those organizations that are considered part of the critical infrastructure, including telecommunications, banking and finance, energy, transportation, and essential government services. Terrorist attacks on the continental United States caused organizations to evaluate ways information can be protected to prevent terrorists from destroying corporate assets or using corporate assets to harm others.

Information security is increasingly the focus of legislation, some of which specifically calls for the creation of information security-focused administrative functions. Electronic information processing and communications in the banking and healthcare industries are now subject to oversight under the Gramm-Leach-Bliley Act (GLBA) and Health Information Portability and Accountability Act (HIPAA) regulations, respectively. Both pieces of legislation specifically require organizations to address information security from the executive level. For example, section 501(b) of GLBA requires covered financial institutions to create management and oversight capabilities that include the formation of security functions, security management responsibilities, and interaction with the Board of Directors. HIPAA requires covered entities to have appropriate "administrative procedures to guard data integrity, confidentiality, and availability." They must implement security management processes that address risk analysis, risk management, and security policy.

Environmental dynamics including those identified above place considerable pressures on organizations to identify appropriately skilled human resources, establish defined and repeatable work tasks, and provide them with tools that enable efficient and effective performance – firms are challenged to develop relevant information security architectures. This chapter will seek to identify and explain important topics related to organizing for information security. We will first examine the current information security landscape of threats and vulnerabilities. We will then review the concept of security as architecture, presenting a framework to describe the organization. Next we will provide guidance on various organizing decisions and identify methods to evaluate the security program. Finally, we will focus on the important areas of leadership as related to information security.

# 2. THE FORCES SHAPING INFORMATION SECURITY TODAY

Many forces drive the ways in which we define and respond to information security requirements. This includes a combination of internal and external factors that appear to be generally increasing in importance, scope, and potential organizational impact.

## 2.1 The Importance of Managing Information

In most organizations, information is a crucial asset, intertwined with business context and opportunity to produce astounding possibilities. Information and environment interact. As described by Davenport [8], the "organizational environment can guide or motivate a particular information environment, just as that information environment can enable or constrain the organization." Increasing attention is being paid by firms in all industries to aspects of information and knowledge management. Identification of information as a critical asset, therefore, provides significant incentive to study information security. Recognition of the importance of environment in promoting or supporting information management provides a glimpse of the vital role, and responsibility, of information security.

The increasing volume and complexity of information or knowledge management activities highlights the need to simultaneously consider security aspects. Barth [4] wrote of the increase in competitive intelligence initiatives, enabled by advances in information technology capabilities and knowledge management philosophies. However, Barth noted, "Whether they're being stolen or simply shared, if your intellectual assets are benefiting your competitors, you aren't managing your knowledge effectively." The essential objectives of information security may be summarized as striving to assure the appropriate confidentially, integrity, and availability of business information. These are activities that may be viewed as complementary, perhaps requisite, to information and knowledge management activities.

Today, information technology is embedded throughout organizations as a key enabler of information management activities. Technology has a profound impact on organizations including the people and processes it serves to integrate. Variances in technology adoption may be visible when comparing vertical industries or geographic markets, however the influence of information technology is quite distinctive.

Rationales for the application of specific information technologies are not always as clear. It may be somewhat obvious that technologies are introduced to organizations to improve information management capabilities with the goal of attaining further benefits such as cost savings or the creation of facilities that allow the firm to reach new markets and offer innovative products. The introduction of information technology is typically justified through a variety of expected benefits such as increased productivity and output. However it is often difficult to measure actual benefits realized. Other benefits often cited are perhaps best described as dimensions of data value or quality including enhanced accuracy, timeliness, integrity, availability, and other factors. The promise of information technology is to enable improved management of information to support business decision making, with the intent of promoting overall organizational success. This promising vision, however, would be incomplete without evaluating a myriad of new management concerns presented by technology and its applications. Therefore, the focused management of technology is necessary to guide in its application, control its related costs, and manage the multitude of new risks created by inserting technology into critical business processes.

The new world of networked, integrated technologies creates significant security requirements. Schneier [24] identified a number of information security needs including the following.

• Privacy. The increased automation of many everyday transactions results in gathering significant quantities of highly personalized information. Standards of responsible management of this information are evolving, with pressures coming from throughout the marketplace including consumers, business partners, and government.

• Multi-Level Security. Information may be classified according to a variety of criteria to identify relative protection requirements. For example, US government labels of "Secret" and "Top Secret" indicate differing information protection requirements.

• Anonymity. All individuals experience some degree of anonymity as members of society. However, the use of computer technologies often eliminates true anonymity due to unique electronic identifications that may be associated with the end user.

• Authentication. Authentication is a requirement in even the simplest of human interactions. The most basic contracts require some level of authentication as to the identities of parties, terms, and consideration.

• Integrity. Modifications (intended or otherwise) to information that occurs following execution of a transaction may result in losses to the end user, and also cause significant damage to the user's perception of the respective information system.

• Audit. Systems require some level of automation to identify usage and determine if system usage was within established standards.

• Electronic Currency. The completion of transactions in the electronic marketplace requires the efficient and reliable transfer of payment through electronic means.

• Proactive Solutions. Systems architects must anticipate and seek to overcome potential shortcomings or flaws in design before the weaknesses are exploited.

Security objectives challenge us to make decisions about information we otherwise seem inclined to ignore. Concerns of individual privacy, ethics, and in some cases morality are perhaps more closely tied in practice to information security than any other dimensions of information management. According to Schultze [25], "The moral value of information depends on distinctly human faculties, such as insight, discernment, and judgment." Similarly, human faculties must be relied upon to effectively define the appropriate management and protection of information. Information technology enables information sharing never before possible, bringing tremendous promise but also clearly presenting society with tremendous potential for misuse. In guaranteeing information security goals such as confidentiality, integrity, and availability, human judgment ultimately forms the basis of security posture. The critical factors that guide and enable the formulation and application of judgment to determine appropriate controls are described in the sections that follow.

## 2.2 Understanding the Dimensions of Risk

Risk may be considered a natural by-product of introducing technology to a business process. While the concept of technical risk may be readily apparent, understanding how technical risk is created may be valuable in helping organizations understand how to effectively respond to and manage new risks. The relationship between business, information technology, and information security may be illustrated using the Gray model of technical risk, below.

The Gray model of technical risk [10] is generally useful to describe how business and technical risk naturally occur as a result of pursuing business opportunities. The model indicates that a business opportunity will present a corresponding set of business requirements. Business requirements drive technical requirements, which are subsequently met with a technical solution. However, the chosen technical solution will inevitably result in the creation of new technical risks.

Business requirements also possess inherent business risks. Such risks are sometimes assumed in the hope of gaining reward. Certain business risks may be reduced by the technical solution. However, a failure of the risk-mitigating technical solution (i.e., a technical risk event coming to fruition) will therefore result in restoration of the business risk in addition to disrupting operations.

An additional important relationship is evident in the Gray model as well; information technology may drive business strategy because a chosen technical solution may enable the creation of new business opportunities. However, any new business opportunities undertaken as a result of the new technology solution will also be subject to the same risk models as the original opportunity. Therefore, mitigation of technical risk becomes essential in building reusable, extensible information architectures that are intended for use as the foundation for future technology solutions and business opportunities. It is apparent that the creation of e-business architecture in practice is typically undertaken with goals of reusability and extensibility.

The relationships between risk concepts serve to underscore the importance of information assurance within organizations. Specific risks visible in practice may be described using the risk taxonomy produced by the US Office of the Comptroller of the Currency (OCC) that highlights four general categories of risk specifically related to the financial services community:

- Transaction Risk: Risk event prevents completion of a business transaction
- Strategic Risk: Risk event impedes ability to execute on strategic plans
- Reputation: Risk event damages organization's creditability
- Compliance: Risk event prevents organization from following regulation

Recognition of a particular aspect of Reputation risk occurred following the events of September 11, 2001, something we may describe as Societal Expectation. Recent events have shown that terrorists present threats to organizations through either (a) directly harming organizational staff and/or assets, and (b) the hijacking of organizational assets to do harm to others. The latter risk came to fruition for United Airlines on September 11. Risks associated with terrorism arguably are associated with all four risk types. However, organizations now are also expected by society (i.e., the marketplace) to take actions to prevent the hijacking of their assets that may result in harm to others. Security over information systems that include detailed flight planning information, for example, may perhaps be considered more important post-September 11 due to heightened sensitivity. Insufficient information security strategy and execution may lead to the creation of such types of risk.

Dimensions of risk may be qualitatively and quantitatively assessed. Technical risk may be evaluated by considering the probabilities of various risk events coming to fruition, and the value or cost of the risk event. Achieving a clear understanding, preferably including quantification, allows the organization to take appropriate mitigating actions.

Risk mitigation tactics include the acceptance, mitigation, or assignment of risk. Acceptance represents an organization's decision to recognize a specific risk and proceed with the intended business strategy without taking mitigation actions, accepting that the risk event may, in fact, occur. Organizations may choose to accept risk where they recognize a low probability of the risk event coming to fruition, or perhaps where the value or cost impact of the event is deemed negligible. Risk mitigation includes attempts to minimize and control for the risk factor. Mitigation tactics may include the addition of systemic or programmatic controls, including the introduction of information security technologies such as firewalls and intrusion detection systems. Finally, organizations may choose to assign risk, or transfer the risk to an underwriter. For example, organizations may purchase any of a number of commercial insurance products that provide coverage in the event technical risk events come to fruition.

Technical risk events appear to vary based on the type of technology employed. Alter [3] identified the following general categories of technical risk events.

- Customer: Operator Error
- Product and Services: Liabilities for System Failures
- Business Processes: Inadequate System Performance
- Participants: Operator Error
- Information: Data Error, Accidental Disclosure
- Technology: Hardware Malfunction, Software Malfunction, Software Design Error, Physical Damage

Alter provided recommendations for action to address the various risk types. Development of quality information systems would certainly reduce the occurrence and impact of risk events. Appropriate user and operator training may help to minimize inappropriate usage of the system, as will the use of strong access controls. Adequate physical security is required to prevent tampering or accidental damage to physical system components. Finally, conducting proactive assessments and audits is vital to identifying potential problems before they occur.

Risk mitigation activities are perhaps best guided by considering the importance of the information processed by the information systems. The relative importance of information may be considered a combination of concerns related to the perceived criticality and sensitivity of the information.

Criticality may be described as the overall value of the information to the continuance of organizational operational processes. For example, the availability of customer information is critical to a manufacturing organization's ability to ship product. A lack of customer information, in this example, would prevent the organization from operating. Inability to ship product represents a transactional risk event and could also have profound effects on the organization's reputation. Further, operational failures may cause the organization to violate contracts or not meet regulatory requirements.

Information may be considered sensitive if, by its nature, public disclosure would represent a threat to successfully conducting business transactions, to the organizations strategy, to the organizations reputation, or to meeting regulatory requirements. Sensitivity must be considered from multiple perspectives including the organization, its customers, or society in general.

Information importance considerations are useful in guiding risk mitigation activities including budgeting. For example, an organization would probably not choose to spend significant sums to protect information that is considered neither important nor sensitive.

It is apparent that information assurance programs must exist within the context of a comprehensive risk management program in order to have meaning and value to the organization. Information protection tactics merely represent activities intended to mitigate risk. Furthermore, we will review later how risk management is a business responsibility and should not be delegated by the business to information technologists. Information technology, however, certainly has a vital role as critical enabler in automating significant pieces of the information assurance program.

## 2.3 An Environment of Threats and Vulnerabilities

The "2002 Computer Crime and Security Survey" report produced by the Computer Security Institute (CSI) and the US Federal Bureau of Investigation (FBI) contained many notable observations [21]. According to the report, ninety percent of organizations were able to detect "computer security breaches" over the past year. Eighty percent admitted they had suffered financial losses due to "computer breaches". Thirty-four percent indicated they had reported suspected intrusions to law enforcement, compared to only sixteen percent in 1996. It is apparent the systems and networks of a large percentage of organizations continue to be breached and in some cases the organization is damaged. Understanding the nature of this phenomenon requires an examination of the threats to and vulnerabilities of our information networks.

It seems that reports of the latest weaknesses in the most popular technical architectures routinely fill airwaves and headlines these days, via both technical and mainstream media. This may be due to a number of factors including increased awareness and sensitivity to security throughout society as well as the ubiquitous presence of information technologies throughout our most vital institutions. Regardless of origination point, it is apparent that reports of technology flaws and architectural deficiencies are no longer confined to arcane Internet message boards and academic computer science labs.

Distributed (or client/server) computing architectures, by their very nature, are built with the intent to achieve networking and internetworking. Distributed technologies enabled the Internet. The distributed computing model is intended to provide interconnecting capabilities. The result is a potentially vast spectrum of technologies working together to achieve a variety of systemic purposes.

While it certainly has enabled radical flexibility in information systems design, and arguably business and market dynamics, distributed computing by its very nature supports and encourages diversity in computing technologies on a wide scale. The resulting installed base of distributed systems, to no surprise, therefore exhibits wide variance in technology architecture and design. An unfortunate effect of this diversity is high levels of complexity when designing and supporting new systems and technologies. Software manufacturers, large and small alike, must try to identify and accommodate a vast array of potential technology platforms, each with its unique mixture of hardware, software, and communications structures.

This complexity makes it virtually impossible for systems designers and builders to identify all possible architectural pre-conditions throughout the installed base of potential end users. The result is a reliance on a phased software implementation approach where systems are alpha and beta tested in controlled settings before release to the general population. End users in the general population subsequently uncover additional flaws, report them to the appropriate vendor, who eventually issue software updates, or patches. This "report and patch" process continues indefinitely while, at the same time, developers incorporate new features, improve performance, or enhance security to design the evolution of the various components of the distributed computing architecture. Refinements in software engineering techniques and patch dissemination utilities should be expected to continue to improve developer's abilities to effectively model architectural pre-conditions. However, it is hard to envision that we will abandon the current software release processes, including the "report and patch" cycle, any time soon.

Now here is where the process becomes even more challenging. While the technology evolution is guided by intelligent, eager, and entrepreneurial designers, it is continuously under attack by an array of mischievous teenagers, criminals, and even terrorists who proactively identify and exploit vulnerabilities presented by aspects of the many, diverse computing architectures we rely on. The software industry has responded to this challenging by leveraging the traditional "report and patch" process cycle as described above, with some notable differences in the source of the "report" and a need for more hastened patch development.

Technical vulnerabilities may be exploited to achieve an attack. However the motives for the attacks represent a separate matter. Schneier [24] grouped cyber-attacks into three general categories – criminal attacks, privacy violations, and publicity attacks. Criminal attacks range from fraudulent transactions to the theft of information to destructive attacks that seek to impede the organization from functioning. Privacy violations are those activities where the attacker seeks to acquire knowledge they would otherwise be prevented from attaining. Publicity attacks are often perpetrated by individuals who attack technology to draw attention to him- or her self, perhaps with the ultimate intent of bringing attention to some political or social cause.

Motives may be analyzed further by considering the locus of the attack's origination. Perpetrators may be generally described as "internal", such as a current or former employee, or "external" where the attack is initiated by someone or group that has never been associated with the subject of the attack in any formal or informal way. Recent studies by CSI have consistently shown large percentages of incidents to be internal, such as the result of actions taken by "disgruntled employees." [21]

The inherent vulnerabilities and design flaws of systems and technologies are under constant attack from a range of threats. The complexity and diversity of our computing landscape should be expected to continue to grow. Organizations must therefore prepare to institute security practices that are well defined and integrated as repeatable processes instead of loosely correlated, reactionary tasks.

## 2.4 The Emerging Landscape of Regulation and Oversight

The Health Information Portability and Accountability Act (HIPAA), authored by the US Department of Health and Human Services (HHS) has dramatically altered the way privacy and security are viewed by the many firms and organizations associated with healthcare industries. Firms that manage protected health information (PHI) are required by HIPAA to exert significant standards of care in information management.

Privacy is a matter of individual capability to exercise control over management of their personal information. The HIPAA Privacy ruling touched on a number of areas important to individuals and businesses alike, and has raised many significant questions of interpretation, including the following examples:

- Patients Review of their own Medical Records. How will the costs of duplicating and distributing copies of medical records as per HIPAA be recovered?
- Safeguards to Protect PHI. Can PHI be transmitted electronically via fax?
- Incidental Disclosures. Are patient sign-in sheets prohibited by HIPAA?
- Business Associates. Does HIPAA require covered entities to monitor business associates with respect to compliance? Who is considered a "business associate"?

Security represents a variety of physical and virtual controls over information, and may be considered an enabler of privacy. The HIPAA security provision is at least equally challenging to understand and apply as the privacy regulation [9, 12]. HIPAA Security requires covered entities to perform assessments to determine their risks and vulnerabilities. Organizations are required to take action to protect PHI against unauthorized access, disclosure, or alteration. HIPAA suggests organizations should "implement and maintain security measures that are appropriate to their needs, capabilities, and circumstances." Clearly there are many aspects of HIPAA left to interpretation by the covered entities, as well as any organizations seeking to assess organizational compliance with the regulation. Interpretation of HHS intent and suggested application with respect to HIPAA has developed into an industry of its own, adding significant complexity to the healthcare landscape.

Just as HIPAA has affected the healthcare community, so too has the Gramm-Leach-Bliley Act (GLBA) changed the way financial services firms approach information security. Perhaps the section of GLBA most related to information security, GLBA Provision 501b became effective in July 2001. In a joint announcement released by the Board of Governors of the Federal Reserve System, Federal Deposit Insurance Corporation, Office of the Comptroller of the Currency, and Office of Thrift Supervision, information security requirements were concisely stated as follows:

The guidelines require financial institutions to establish an information security program to: (1) identify and assess the risks that may threaten customer information; (2) develop a written plan containing policies and procedures to manage and control these risks; (3) implement and test the plan; and (4) adjust the plan on a continuing basis to account for changes in technology, the sensitivity of customer information, and internal or external threats to information security. Each institution may implement a security program appropriate to its size and complexity and the nature and scope of its operations. The guidelines outline specific security measures that institutions should consider in implementing a security program. A financial institution must adopt those security measures determined to be appropriate.

The guidelines also outline responsibilities of directors of financial institutions in overseeing the protection of customer information. The board of directors should oversee an institution's efforts to develop, implement, and maintain an effective information security program and approve written information security policies and programs.

The guidelines require financial institutions to oversee their service provider arrangements in order to protect the security of customer information maintained or processed by service providers. Each institution must exercise due diligence in selecting its service providers, and require its service providers by contract to implement security measures that safeguard customer information. Where indicated by an institution's risk assessment, the institution must also monitor its service providers by reviewing audits, summaries of test results, or other equivalent evaluation of its service providers, to confirm that they have satisfied their contractual obligations.

GLBA addressed many topics under the domain of information security management. The regulation called for the creation of a security program, suggests specific program components, identifies the board of directors as playing a key role in program development and management, and requires banks to provide oversight to outsourced service providers such as application service providers (ASP).

The Basel II Accord represented a significant step forward on information security in the international banking community [5]. Drafted by the Bank for International Settlements (BIS), Basel II made strides in the description of operational risks faced by global banks and offered recommendations for continuous risk management. Basel II defined operational risk as "the risk of direct or indirect loss resulting from inadequate or failed internal processes, people, and systems or from external events." Basel II effectively identified operational risk as a significant management concern for the world's largest banks and offered specific recommendations to support continuous risk management activities.

The draft release of the US National Strategy to Secure Cyberspace similarly included recommendations for addressing emerging technical threats and vulnerabilities across various segments of the population. The recommendations relating to large enterprises touched on many popular themes including suggesting the need for more involvement from leadership including management and board members. The draft specifically recommended that large enterprises establish corporate security councils. While particularly vague on points of enforcement, the draft provided a glimpse of what we should expect to emerge over the next few years in the form of more meaningful (and probably legislated) requirements.

Apparent throughout much of the recent legislation and industry commentary on security is a common theme that vertical industries and the organizations within them need to exercise appropriate judgment in the course of assessment, mitigation, and oversight. In other words, the recommendations from many sources point to the identification of, and compliance with, some notion of reasonable care as defined by the firm or industry. A recent review of topics of information law, for example, concluded by stating "companies will need to exercise care when they place their data on open networks, the routings of which they can not control.". The emergence of information sharing committees and government involvement in information security represents in some ways institutionalized actions intended to define reasonable care. We should expect future regulation, and perhaps litigation, to focus significant attention on definition, interpretation, and compliance with standards of reasonable care.

## 2.5 The Organizational Challenge

Many technical, process, and human factors combine to create concerns that are generally described as within the scope of information security. Organizations should be expected to continue to build information systems and integrate technologies that present profound technical vulnerabilities. Businesses will continue to deploy them to achieve efficiencies and competitive advantages despite the shortcomings of the underlying tools. Therefore, effort to understand, mitigate, and control inherent technical risks must be supported through a variety of activities. Risk analysis, mitigation, and the continued maintenance of secured operations require the creation of an appropriate information security function that is tailored to the unique challenges encountered by each firm. Adding further complexity to these relationships is a shifting vulnerability landscape that requires continuous monitoring and adjustment via appropriately skilled staff, well-defined organizational structures, and well-defined repeatable processes.

There are a variety of approaches in practice to organize for security. In the following sections we will consider key organizational constructs that are important components of information security and present corresponding organizational recommendations to help solve the problems of managing technical risk in practice.

# 3. DESCRIBING THE INFORMATION SECURITY ORGANIZATION

Analysis of the challenge of organizing for security requires a basis of understanding of core organizational constructs. The following is a review of important constructs and commentary on how they may be considered in an evaluation of information security.

## 3.1 Information Security Architecture

Architectural models are useful to illustrate how domains of any larger whole combine to form a greater, complex structure. In other words, architectural models enable and support analysis. Academic and practical models exist that provide a glimpse of the breadth and depth of activities normally associated with information security. Murray, for example, provided an Enterprise Security Architecture model intended to account for a variety of technical requirements, services, and functional objectives [17]. The following summary draws from a variety of sources to provide a more comprehensive overview of security as architecture, including human resource, process, and tool domains. In this example, the individual subject domains may each be considered "architectures in the small", while the overall structure they combine to create may be described as "architecture in the large". We will adopt this modular representation to analyze information security in the organizational context, considering dimensions of human resources, processes, and tools.

### Human Resources Domain

The human resources architecture represents the domain of people. Many important characteristics emerge when examining this domain including technical skills requirements, business and organizational acumen, and organizational structure such as the formality of roles and responsibilities. Additional concerns extend beyond the security organization into the general employee population, such as information security awareness.

The technical skills requirements of the information security staff appears to be in constant debate, with otherwise similar organizations choosing very different paths with respect to technical prowess. Anecdotal observations indicate that organizational size may influence degree of socialization, as larger organizations tend to require information security oversight to guide large, distributed technical staffs. Regardless of degree of skill level, it is imperative that information security personnel understand all elements of the respective information technology architecture and its internal technical linkages to be effective at all in developing appropriate security frameworks.

A strong appreciation of business and organizational topics is required as well. Organizations may be considered products of many influences including their respective social context and history. Information security programs must react to the continuous changes of organizational characteristics; therefore security personnel must continuously monitor the internal environment. We will address a number of additional concerns related to the human resources domains as defined by organizational theory and behaviour in later sections.

Firms today may also choose to outsource all or significant portions of the information security organization. Outsourcing may provide a number of significant benefits including the shifting of hiring and staff education to the outsourcer. Firms with relatively low levels of on-staff information security expertise can use outsourcers to quickly address critical problems while theoretically providing the internal staff with opportunities to learn from the outsourcer. A significant challenge to the outsourcer, however, is the requirement to understand the organizational context of the subject firm. Knowledge of company structures including core processes as well as culture must often be supplemented by close supervision or involvement of the internal staff.

## Process Domain

The essence of the information security function is represented in the process domain. The process architecture is one of the more developed areas of study in information security. A number of existing models provide a glimpse of the way organizations tend to define the functional objectives of information security.

The Certified Information Systems Security Professional (CISSP) designation has emerged as a leading indicator of professional competence on a wide variety of topics related to information security. The CISSP structure presents a process-focus to define the essence of information security. Administered by the International Information Systems Security Certification Consortium (ISC2), the CISSP framework divides security into the following ten domains: [14]

- Security Management Practices
- Access Control Systems
- Telecommunications and Network Security
- Cryptography
- Security Architecture and Models
- Operations Security
- Applications and Systems Development
- Business Continuity Planning and Disaster Recovery Planning

- Law, Investigations, and Ethics
- Physical Security

Treese & Stewart [29] also present a process-centric view of security by describing the steps taken to "build-in" security into the environment including the following sequential phases:

- Evaluate
- Define Policy
- Design Environment Security
- Design Application Security
- Monitor
- Provide Feedback to Support Evaluation, and Repeat the Iterative Process

Examining the critical domains of protection, detection, response, and governance may help us develop a generic architecture for information security processes. Protection activities represent largely proactive planning, preparation, and communication efforts. Detection includes operational monitoring of technical and non-technical indicators of information protection. Response actions are taken as a result of observations or detections. Finally, governance activities may be described as oversight practices. Schneier [24] observed that in practice, "Digital security tends to rely wholly on prevention: cryptography, firewalls, and so forth. There's generally no detection, and there's almost never any response or auditing." My own experiences as an information security consultant are consistent with Schneier's observations. It is also apparent that little attention has been paid to formal governance processes. The following generic architectural domains are intended to provide a more complete framework to guide the design and organization of security processes.

## Protection Processes

Information security planning may extend to define the roles and functions of all architectural elements including people, processes, and tools. Planning should be supported by iterative cycles of risk assessment and requirements planning. Information security planning is typically based on assessments of the importance of information and is concerned with guarding information as a critical asset. Network security represents a supporting function that is concerned with the technical protection measures necessary to protect information. Therefore, network security planning should occur subsequently and as a complement to information security.

Application security is approached in a similar way; the value (and therefore) risk associated with the application is determined largely by the importance (criticality and sensitivity) of the information being processed. Information security planning must clearly be accomplished by considering information value first which in turn will establish the relative importance of technical security aspects. One of the most important proactive steps in the information security program may be generally described as awareness activities. Information on technical risks, threats, vulnerabilities, and associated counter-measures may be provided throughout the organization to improve overall security awareness. Attention to awareness should be anticipated to encourage all staff, not just information security personnel, to consider information security while performing their respective functions. Awareness training in some organizations may extend into the user community.

## Detection Processes

Detection activities within organizations should include people, processes, and tools organized and chartered to discover suspicious computing-related activities. Human investigators play the most critical role, as they must use whatever means at their disposal to alert themselves to potential issues. A variety of monitoring technologies exist to assist the investigator in scanning the environment. Detection processes logically seem vital to effective monitoring of overall information security effectiveness. However, it is apparent that many organizations also have inadequate tools and poorly trained staff in monitoring and detection roles.

## Response Processes

Response steps include all actions taken after observation of some sort of "trigger", or output of the detection process. Response tasks include everything from using technologies to capture electronic observations, to the completion of technical investigations and forensics, as well as physical investigation. Response complexity should be expected to increase with organizational size, as large organizations often spread detection and response activities across multiple individuals and groups. Such specialized structures may create increased reliance on process integration and, as a result, increased process formality and structure.

## Governance Processes

Oversight and governance are often overlooked in information security programs but may represent critical components. At a high level, governance tasks may be divided into planning and operational categories. Planning tasks associated with governance may include the development of policies and procedures, establishment of measurement criteria for security effectiveness, the creation of security budgets, and oversight of the operational risk management program.

While presenting on the topic of information security policies and procedures to the CIO and IT management team at a Global 1000 organization, I was interrupted by a senior IT director whose impatience was noticeably growing throughout the discussion. The IT director interrupted the presentation by exclaiming, "that since many other organizations had undoubtedly encountered, and probably solved, the same problem we were discussing by producing a policy" that I should simply state the price at which I would re-sell the same policy to him, regardless of the unique context and nature of his organization. This exchange illustrates some of the most troublesome issues and challenges encountered when developing sound, helpful and appropriate policies and procedures that are truly meaningful to the organization they are to be implemented in. In many ways, policies and procedures represent the "glue" that binds the security architecture.

## Tools Domain

It is apparent that the vast majority of attention paid to information security topics to date has emphasized the domain of technical tools. Information security technologies have evolved rapidly as distributed computing and commercial reliance on the Internet has exploded. While this chapter will not address tools directly, it is important to note the existence of a baseline set of information technologies is rapidly becoming expected as part of an organization's "reasonable care" of it's information assets. Technologies such as anti-virus scanners, firewalls, intrusion detection systems, and vulnerability scanners appear to increasingly be considered as baseline elements of the technical architecture—this was simply not the case until very recently. Organizations should anticipate seeing this technology baseline increase as the threshold for providing "reasonable care" should be expected to continue increasing and, perhaps, be codified in new regulation and legislation.

## 3.2 Concepts of Organizational Theory

Organizational theory literature provides a description of critical characteristics that may be used to describe the essence of organizations. These characteristics can be extremely useful in analyzing organizational requirements for information security. The literature provides insights on what can be described as core organizational constructs including the following.

### Environment

The environmental context provides many important drivers of security. The internal environment encapsulates a number of important security drivers including information and the systems that process it. Dimensions of organizational culture such as how risk is considered are included here as well. The internal environment also includes all aspects of the corporate security program. Similarly, the external environment represents a significant source of threats. The overall importance of information security increased dramatically as organizations established external connectivity via the Internet – the Internet linkage established significantly greater interaction with the external environment, and therefore altered the risk framework. Regulatory bodies are increasingly becoming part of the external information security environment for many organizations as well.

### Power and Control

Organizational power and control structures are important considerations in many problems involving the management of technology and related risk. As stated by Davenport [1997], "Information is not innocent". If information truly is power, information security personnel find themselves in quite a conundrum; they may be perceived as disruptive to established power and control structures while legitimately executing their charter. The control of access to information is a highly political concern in most organizations. Information security professionals must continuously weigh political concerns and, therefore, must be effective in balancing the sometimes competing interests of information sharing and information protection.

### Strategy

The formulation and execution of strategy provides direction and prioritization to the information security program. Established corporate objectives (IT and business) provide significantly valuable insights to guide the determination of information importance and, as a result, information protection approaches. From another perspective, the information security program itself represents an important element of the organizational strategic plan. By considering information security a strategic interest, the need to establish appropriate organizational involvement and representation should be clear.

### Size

Organizational size is examined in many studies of organizational theory and evaluated as an influencer of various practical observations. In the security context, it appears that size may influence a variety of characteristics related to organizing. For example, as mentioned earlier smaller organizations may choose to have more technical security staffs and assign hands-on technical duties to those staffs, while many larger organizations appear to view security as an advisory and oversight function. The effects of size on security organizations represent an important area of consideration.

### Process

The generic process architecture described previously illustrates the importance the concept of process plays in defining information security functions. The utility of a security program typically appears to be a function of its process performance in areas relevant to the subject organization. Processes may occur at varying degrees of formality, with similar disparity in the level of sponsorship received from management.

### Function

Closely related to process is the idea of function. The goal of division of labour requires specialization into a number of organizational functions. Security in a given organization is typically defined based on specific activities or functions, prepared for or undertaken within the broad category of information security. The definition and degree of specialization in information security should be expected to vary by company. For example, as mentioned previously, differences in organizational size may lead some organizations to build security staffs that are more or less technical. Similarly, some organizations seek to promote high levels of specialization of the information security function while others choose to mix information security tasks with other IT operational activities.

## Structure

Literature on organizational theory provides ways to describe organizational design using a variety of characteristics. "An organization's structure defines how it is organized, usually by means of some 'formal' coordination mechanism" [11]. Formalization refers to the prevalence of structured, often documented, organizing rules. Complexity characterizes the degree of task differentiation. Centralization refers to the characteristics of the "typical" decision making process with respect to the locus of decision-making. Hierarchical reporting structures are another important structural characteristic. It is apparent that in practice organizations take sometimes widely different approaches to defining information security organization characteristics such as span of control, organizational level, and choosing between functional and matrix structures.

## People

The role of people in information security endeavours should be quite apparent. Information security requires an integration of knowledge domains including business, organizational, and multiple technical disciplines. Although significant, fundamental challenges have been overcome by many different new information technologies, the ultimate planner, investigator, and actor on security matters simply must be a broadly-educated staff person who will be able to review varying levels and types of information to reach some set of conclusions or recommendations. Significant areas of information security assessment, planning, and investigations require dedicated human resources.

From a different perspective, people also represent the most significant threats to information. Human actors pursue information resources for a variety of motives. They may threaten the organization as outsiders, or may work internally as employees. They may come from a diverse group of political, social, and ethnic groupings. It is safe to say a person is behind every intentional "misuse" of an information technology. Therefore, the design of our information technologies and information security structures should be accomplished with a motivated, human "attacker" in mind. For example, a human attacker will act according to a specific motive to access valuable information resources. Therefore, it is by studying anticipated motivations that information security planners may determine which resources represent the most likely targets. The human dimension of security may be the most critical one, including those working from perspectives of "attacker" or of "protector".

## Information

Although we sometimes blend the concepts, information exists separately from information technology architectures. I'm quite sure that many organizations would describe information as one of the essential ingredients to their success. However, I also would not be surprised if those same organizations had difficulty explaining just what information is, and where it is located. Information may be viewed according to a continuum of data, information, and knowledge. Data represents discrete facts, or observed states of the world. Information was described by Peter Drucker as "data with relevance and purpose". Knowledge may be considered a largely internal concept, with each person adding personal meaning, and therefore increased richness. By considering these dimensions, it is easy to see that information, in all of its forms, is prevalent throughout organizations. Information Architectures, therefore, include digital and non-digital structures. It is apparent that in most organizations, information security personnel focus almost exclusively on digital information resources and ignore other forms.

## Tools

Electronic tools and technologies for information security were described earlier. However, the organizational concept of tools extends beyond the digital information technology toolset. Any device or structure that assists in the transformation of input resources to output products may be described as a tool or technology. In practice, organizations may be led to focus protection on critical databases used by business applications, and not emphasize other digital structures such as business models that may be of potentially greater value. There are a variety of tools used within organizations in the completion of various business functions. For example, a complex pricing model used by a money management firm would be described as a tool. Such a tool would represent a significant business asset that should be protected. The pricing model itself does not represent information. However, a specification of it does represent information (and arguably knowledge), and may be stored in some data base or file server. Therefore, the importance of internal tools must be considered when establishing protection goals.

## 3.3 Concepts of Organizational Behavior

Organizational Theory is useful to show us what organizations are. Concepts of organizational behaviour provide important insights into how people interact within the organizational setting. A variety of factors that may influence the application of information security are apparent in reviewing organizational behaviour literature including the following.

### Personality

All managers must account for variances in personalities within their teams. In the realm of information security, managers are asked to assemble a cross-functional team, with varying levels of formality with respect to roles and responsibilities, including technical and non-technical staff. Because of the need to base information security decisions on the output of sound risk management activities, managers must bring together (or at least communicate with) an extremely diverse group that will most likely include business and information technology staff. Non-technical staff will be challenged on tools and technologies topics, while technical staff will be forced to consider business and organizational impacts of their system designs and operational processes.

### Motivation

Individual motivations can vary widely and there are a number of models that try to explain the complex dynamics of motivation. From the highest level, managers must expect to observe a variety of motivational forces and be prepared to respond to them while shaping the information security function. For example, an application development staff may be under significant pressure to deliver a project by a certain date, perhaps even be offered a bonus for achieving the milestone. Suggestion of additional review steps to assure information protection may not be popular in such a situation! How then, can an information security manager compete with the project schedule and the business needs that drive them?

### Stress

The above example of the development team infers another important behavioural dimension—stress. Individuals are provided with a variety of positive and negative motivators during the completion of any project. The accumulation of stress during these activities may cause topics such as information security to be considered of lower priority than meeting functional requirements. Stress may also come to play in the face of unknowns, or changing circumstances. For example, systems recognized as part of the US critical infrastructure may be potential targets for internal terrorists. This condition was not fully recognized prior to September 11, 2001. Systems designers and developers are now under significant stress to quickly reconsider past assumptions to ensure information protection.

## Group Dynamics And Teamwork

Many examples and comments provided earlier point to the challenges of developing a cross-functional team to effectively address information security. Management of such teams can be extremely challenging, particularly when the teams are established with minimal formality. I have seen many managers assemble information security teams, but only a few have concerned themselves with the inter-group processes that will determine the success or failure of the mission.

## Communication

Generally speaking, technologists are not often commended for their communication practices, particularly in sharing knowledge with non-technical audiences. Similarly, business and management staff do not always take the time to articulate the important factors that should drive technical activities. Communication between all participants in the security function is essential and is not resolved by relying on electronic mail! To be effective, there must be genuine dialogue established within the cross-functional team.

## Leadership

Leadership will be addressed in greater detail in a subsequent section as it represents arguably one of the most significant dimensions of the security organization. Leadership has been defined in many different ways, although common themes emerge. At the highest level, leaders serve to provide a picture of a future state, and to move teams towards that state. Their actions send very powerful messages to all team members, intentionally or otherwise.

## 3.4 Conclusions on Organizational Theory and Behavior

So far we have reviewed a number of models that help illustrate what security is and how we may view information security in the organization from a number of perspectives. All of this information provides valuable insights for organizations attempting to design, or redesign, their information security organization.

Outwardly similar organizations, though, have produced very different information security organizations. We can generally describe the possible organizational structures as centralized, decentralized, and federated. Centralized structures are characterized by the consolidation of information security expertise and decision making into a core group that is given broad authority to enact security throughout the organization. In contrast, decentralized structures are those where information security roles and responsibilities are spread throughout the organization and provide little direction from any single group. Decentralized organizations, rather, allow individual functions or business units to set security strategy. Finally, the federated structure represents a hybrid solution where there is a core security team that sets policy, but execution of policy is carried out by staff that are dedicated to specific functions or business units. Is there a single best answer to the question of how to structure the information function – clearly not. However, considering the dimensions covered earlier that require significant business input and direction, it is arguable that decentralized or federated structures may foster improved alignment with business functions and, therefore, result in more meaningful and effective information security. However, unique features such as culture should lead organizations to choose the structure appropriate for them.

Perhaps one of the best approaches to determining, implementing, and continually evaluating the information security organization is to institute a process of assessment using both internal and external resources. Information security assessments can provide important evaluations of the information security program with respect to the business and organizational context. The organizational aspect of the assessment should address the areas addressed previously such as security roles and responsibilities, protection processes, detection processes, response processes, governance, and leadership. A variety of assessment approaches exist that may provide topical guidance and promote consistent standards, including the following.

## National Security Agency's (NSA) Information Assurance Methodology (IAM).

IAM is a robust information security methodology that is often used by external consultants. IAM provides a set of structured assessment categories to examine information protection categories. It is based on a firm understanding of information and system criticality that should drive the assessment program. The NSA regularly offers training and education programs on IAM.

### The Software Engineering Institute (SEI) OCTAVESM Approach

Designed specifically as a self-directed assessment methodology, OCTAVEsm provides organizations with a complete set of process guidelines to evaluate and manage risk, implement OCTAVEsm in an organization, and ultimately develop a protection strategy that is tailored to the organization [2].

There are a variety of reasons for organizations to conduct assessments using both internal and external resources. The use of external resources clearly reduces concerns of "conflict of interest" that may arise from assigning internal staff to assess or audit other internal staff. External assessors also may provide additional knowledge to you that may have been acquired by working with a large number of organizations. In some cases, the use of external resources may be driven simply by a lack of adequately trained internal resources. However, organizations may only progress in the development and refinement of their information security structure through the use of internal assessments that place pressure on internal staff to better understand their own environment. Knowledge acquired during self-assessment may prove valuable during the completion of security operations, as staff members will be better prepared to understand the mission and execution of various steps in the security program.

The study of organizational theory and behaviour has produced a very rich resource that describes the essence of organizations and the people within them. This robust body of knowledge offers significant insights on the topics of creating and managing the information security function. As the prior sections described, creation of the security program should be accomplished in a way that is consistent with the overall mission of the organization and all of its constructs. Without adopting such a comprehensive viewpoint, the planners of the information security function may not be able to craft a valuable and relevant information program. The following section will address the requirements of organizational leaders as they develop the information security architecture.

# 4. A SEARCH FOR LEADERSHIP

Leadership is the attitude and motivation to examine and manage culture – Edgar Schein [23].

The descriptions of core information security functions reviewed previously infer a strong reliance on senior leaders to establish direction and vision with respect to risk management and information security program design. The involvement of senior leaders, in business, information technology, and information security functions is required for the information security program to be effective.

## 4.1 Relationship of CIO and CISO

The two most visible leadership roles related to information security in most organizations are the Chief Information Officer (CIO) and Chief Information Security Officer (CISO). The CIO role has been the subject of numerous academic studies, providing us with a sound, although generalized, appreciation of its major goals and expectations. Quite the opposite, the CISO role is a fairly recent phenomenon and while implemented in a variety of organizations, its essence appears to be subject to significant interpretation. There is apparently wide variance on many important characteristics of the CISO job across organizations. The following is a review of general characteristics of the CIO, CISO, and their common interactions based on a variety of sources including my own practical observations.

### Role of the CIO

The Chief Information Officer (CIO) role appears to have become commonplace in many organizations and in many industries. While each organization views the roles and responsibilities of their CIO in a relatively unique way, there appears to be some common objectives of all CIO's. CIO Magazine recently published an excerpt of a Gartner Group paper that described the responsibilities of the CIO as follows:

- Business technology planning process — sponsor collaborative planning processes
- Applications development — new and existing for enterprise initiatives and overall coordination for SBU/divisional initiatives
- IT infrastructure and architecture (e.g., computers and networks) — running as well as ensuring ongoing investments are made

- Sourcing — make vs. buy decisions relative to outsourcing vs. in-house provisioning of IT services and skills
- Partnerships — establishing strategic relationships with key IT suppliers and consultants
- Technology transfer — provide enabling technologies that make it easier for customers and suppliers to do business with the enterprise as well as increase revenue and profitability
- Customer satisfaction — interact with internal and external clients to ensure continuous customer satisfaction
- Training — provide training for all IT users to ensure productive use of existing and new systems.

As the title implies, CIO's are expected to manage the organization's information and/or knowledge including the development of enabling automation, or information systems. This includes responsibilities at various stages of systems development activities including interaction with the business (i.e. socialization intended to develop shared meaning between IT and business), requirements definition (i.e. externalization to formalize user/business needs), solutions development (i.e. combination of business and technology knowledge), and integration of the solution with business and operational processes (i.e. internalization of the solution). CIO's may be expected to show leadership during each of these general stages.

Similarly, the CIO may be considered generally responsible for introducing capabilities to enable or enhance information/knowledge management. Effectiveness in leading the organization through the development stages may ultimately be judged by the perceived value of the resulting information system's functionality. Therefore, it appears reasonable to conclude that CIO's are rewarded for delivering business-appropriate information management capabilities and should be expected to focus on those activities in practice.

## Role of the CISO

The Chief Information Security Officer (CISO) position takes on a variety of roles and responsibilities in practice. Much of the following commentary is based on my personal observations made working directly with CISO's. As any leader, the essential duty of a CISO may be described as the charter to establish an organizational vision with respect to information security. The CISO appears to be expected to deliver organizational leadership on at least the following major tasks:

- Development of the information security architecture, which includes the integration of people, processes, and technologies
- Protection of the organizations intellectual capital or information assets
- Privacy advocate for customers, employees, and business partners
- Business domain expertise to enable organizational understanding of technical and business risks
- Ability to translate technical jargon into dialogue that is meaningful to the CIO and to the business, to enable business/IT strategic alignment
- Inventory and assess knowledge assets
- Establishment of operational processes to maintain adequate protection and adjust to emerging threats and vulnerabilities
- Promotion of technical risk management practices
- Educator on Information Security topics across IT and business

Organizations approach the role of CISO with varying degrees of formality. This variance does not appear to be directly correlated to organizational size. Another characteristic that appears to vary is the degree of CISO involvement with the business.

Recent CISO hires by a small sample of firms indicate a tendency towards significant levels of experience, technical skill, business acumen, communication skills, and professional designations in information security and audit.

Perhaps the most significant role played by the CISO is that of Risk Manager. As described previously, the management of risk that results from the selection of specific business strategies (i.e. business risk) or information technology solutions (i.e. technical risk) should be of significant concern to organizational leadership. Business or technical risk possibilities that come to fruition often (perhaps typically) result in monetary losses. Therefore, it may be argued that all types of risks to organizations ultimately represent financial risks. Regardless of type, risks are typically taken (i.e. risk acceptance or attempted mitigation) when the organization believes they can accept or mitigate the known risks to such a level that the organization reasonably expects to (a) avoid the possible outcomes completely, or (b) achieve a level of reward or return for taking a specific action that presents the risk.

Understanding and quantifying risks presented by information technology has proven to be problematic. As is the case with most intangible assets, the valuation of information, for example, may be closer to art than science. The understanding and quantification of risk represents an important challenge for the CISO. In practice, organizations may dismiss information security concerns as they perceive the chances of a technical risk "coming to fruition" to be low, perhaps because system intrusions are notoriously underreported by other companies. This scepticism may be exacerbated by a failure to quantify potential loss numbers. An organization's reputation may be shattered as the result of a system breach however the valuation of that reputation is not easily accomplished. Therefore, it becomes quite challenging to request resources to mitigate technical risks because "return on investment" for risk mitigation tactics is not easily calculated.

Risk and Risk Management are important concepts that fall under the domain of all organizational leaders. However, risk related to information systems (including business and technical risk) appears to be increasingly recognized as crucial to organizational viability. Similarly, first-hand observation indicates the management of risk related to information systems is typically considered the responsibility of the CISO.

## CIO and CISO Interaction

A useful exercise to highlight the similarities, differences, and interactions of the CIO and CISO is to consider a "typical" information systems organization where the CIO and CISO collaborate to produce secure information systems. In practice, the CIO organization seeks to maximize the business value of information technology by delivering systems with rich functionality that is relevant to the business. The CISO typically acts to ensure development of a secure system—a fundamentally different goal.

Such an example brings to light a critical philosophical question; a potential conflict of interest in organizations where the CISO reports through the CIO organization. Simply stated, if the CIO is rewarded primarily for delivering information management capabilities then it may not be in the organization's best interest to include the CISO within the CIO organization. An alternative structure would be placement of the CISO in a unique organization or perhaps reporting through the finance/audit function. If they report through the CIO, the CISO may be not be politically able to provide reasonably open and honest commentary and observation if doing so may damage the reputation of their superior (the CIO) or their functional organization (IT). Taking the appropriate actions with respect to information security is not always popular with end users who demand functionality as promised on project timelines, and who may not fully understand the relevant risk factors.

From another perspective, if an organization chooses to place the CISO within the finance or audit function, this may influence CISO skills and qualifications as the person may be selected on the basis of ability to integrate with finance/audit staff rather than technical expertise. In other words, the decision of where the CISO role should be placed may influence the roles and responsibilities, and therefore the skillset, of the ideal CISO for that specific context. This in turn may influence the CISO's methods in building information security architectures.

The same factors may influence aspects of organization structure. For example, technically oriented CISO's may need less support from IT "subject matter experts" than a non-technical CISO. Therefore, the selection of a non-technical CISO may result in increased reliance on matrix structures to allow the CISO to tap into the existing technical knowledge base. The example and similar points made above produce a number of potentially important analysis topics such as the following:

- The inherent conflict of interest between the CIO and CISO, and evaluation of the practical usefulness of approaches of "self assessment" or auditing.
- Approaches to planning security architectures, and how those plans integrate with IT plans (including technical and financial aspects).
- The apparent reality that although users don't always ask for information assurance (either because they don't understand the problem or perhaps are assuming IT will take care of it) they do expect it.
- CIO's may assume programmers will "build in security" although most haven't been trained to build secure systems. Programmers are also often over-allocated and placed under significant pressures to produce functionality; security often takes a back seat. Turnover and the use of consultants also means many programmers will be gone soon after the system is built and therefore there may not be adequate motivation for them to build secure systems.
- CIO organizations may seriously consider or accept the biased advice of product vendors or systems integrators who assure them that "security is built into the system". Information security is context-sensitive. Therefore, the "information security fitness" of a specific technology (i.e. application software) is of minimal importance with respect to achieving adequate levels of information security within a given organization.

- The CIO or CIO organization may not understand vulnerabilities or technical exposures related to their architectures, which in fact change every day.

It appears that the role of the CISO may be, in some ways, similar to the process role of the CIO, however with a somewhat different mission. The CIO is responsible for managing information through the development of information architectures that are aligned with the business, while the CISO is charged with information protection though development of a robust security posture and for aligning the security posture with the business. General control principles suggest the separation of these duties in practice.

The following are a number of questions related to the CIO-CISO relationship that may help organizations plan the CISO function. Answers to these questions seem to vary significantly across the organizations that I have observed.

- What does the CISO consider their charter to be? Do they emphasize the comprehensive assurance of information or do they focus on technical aspects (i.e. information assurance as defined by the NSA vs. technical computer security)?
- How does the CISO align information security architecture development with other IT activities?
- How does the CISO align with the business? Do they interact directly with the business or do they tend to work only through IT? How do they interact with e-commerce teams or initiatives? (i.e. do they influence architectures or do they retro-fit security after e-business architectures are designed?)
- Who does the CISO report to? What organizational level are they?
- Where does the CISO organization report? Does the CISO work within IT, a dedicated security organization, or internal audit?
- Does the CISO have dedicated staff or are they matrix managers? Are security responsibilities centralized or dispersed?
- How high is the CISO's exposure in the organization? Board? CEO? CIO?
- What are their professional qualifications? CISSP? CISA? CPA? How technical are they?
- Does the CISO officially have the title of CISO, and/or do they have de-facto ownership of security along with holding other responsibilities?
- What is the CISO's leadership style?
- Is the CISO asked to interact with (or manage) physical security personnel and activities?

## 4.2 Emerging Organizational Leadership Roles

The dimension of leadership with respect to information security continues to change at a rapid pace. More and more, we see information security topics attaining "top billing" in business publications and even the mainstream news media. New legislation related to information security specifically identifies the need for active leadership from business and information technology areas. As described by Schein [1992], leadership represents activities related to building culture. Leaders today are being asked to build cultures of information security within their existing organizations – they are being asked to lead organizational change.

The leadership requirements for a given organization at any specific time are related to factors such as the present stage of organizational development (i.e. organizational lifecycle stage) as well as any specific challenges the firm is currently facing. Therefore, the decision of how to best achieve senior leadership of the information security function will vary by organization. Despite the implementation differences, it is probably safe to conclude that most organizations in today's environment will need to involve senior business leaders such as the CEO, COO, CFO, and board in the activities of technical risk management and information protection.

Specific actions that organizations may take to raise the profile of information security include increasing board of directors' involvement by placing security topics on the board's agenda. Some firms include a recurring update or status report on information security at all board meetings. Organizational risk management policies and procedures may be communicated to business leadership via the board. Further, senior managers are also in the unique position of being able to define the true leaders of information security, provide support, and establish accountability for performance. Board and management leaders may also evaluate the information security program by using external consultants to critique and perhaps test the information security program. Finally, the creation of an executive steering committee that is chartered to provide oversight for information security functions would establish a permanent agenda item and, therefore, linkage between business and information technology management on security topics.

# 5. CONCLUSIONS

Information security, as described within this chapter, is clearly not a subject that may be addressed with technical solutions alone. Business needs or objectives tend to drive the adoption of new technologies that, in turn, result in increased risk. Organizational understanding and mitigation of such risks requires involvement from parties involved in the management of the foundational business requirements to be effective and non-disruptive. All organizations are called to relatively evaluate the importance of information and underlying information systems, and the evaluation activities absolutely require some level of business and management involvement. The policies and procedures that guide such evaluations are perhaps best designed by those at the highest levels of the organization.

Each organization is therefore faced with the requirement to craft a security posture or management philosophy. The posture each firm chooses to take will vary as a result of differences in each organizational construct. Firms are faced with the need to first identify their appropriate security posture, and subsequently implement a relevant architecture that includes domains of human resources, processes, and tools. Process functions of protection, detection, response and governance to a great extent define the essence of the organizational architecture for security.

Each organization will be faced with unique challenges in implementing the security organization. A number of general concerns, however, may be helpful to consider. Tudor [30] referenced the following factors that may impede a firm's ability to implement a security program.

- Lack of Executive Management Support
- Executive Management's Lack of Understanding of Risks
- Insufficient Resources
- Impacts of Mergers and Acquisitions
- Independent Functions Across Business Units
- Conflict of Mainframe & Distributed Models
- Corporate Culture
- Lack of Understanding of Controls Requirements
- Third Party Management of Technologies
- Rate of Technology Change

Although by no means should we consider it complete, a rich body of knowledge exists on the study of the organization. Significant contributions from a variety of disciplines help us to understand essential organizational constructs including human interactions. However, the study of information security from a management perspective is just beginning. While significant trends such as the formalization of CISO roles are apparent in many organizations, we can assume other important organizational changes are underway as well. The suspected changes should be examined by both academia and industry to provide greater insights on effectively managing information security in organizations.

## REFERENCES

1. Anonymous, 2001. Information Security Oversight – Essential Board Practices. Board Leadership Series. National Association of Corporate Directors.
2. Alberts, Christopher & Dorofee, Audrey 2003. Managing Information Security Risks – The OCTAVEsm Approach. New Jersey: Pearson Education.
3. Alter, Steven 2002. E-Business Security and Control. Information Systems – The Foundation of E-Business. New Jersey: Prentice-Hall. pp. 510-549.
4. Barth, Steve 2001. Protecting the Knowledge Enterprise. Knowledge Management, March 2001. New York: Freedom Communications. pp. 44-52.
5. Basel Committee on Banking Supervision 2001. Consultative Document – Operational Risk. Bank for International Settlements. www.bis.org.
6. Brown, David B. 1995. Technimanagement – The Human Side of the Technical Organization. New Jersey: Prentice-Hall.
7. Daft, Richard L. 1998. Organization Theory and Design. Cincinnati: South-Western College Publishing.
8. Davenport, Thomas H. 1997. Information Ecology – Mastering the Information and Knowledge Environment. Oxford: Oxford University Press.
9. Fox, S. & Wilson, R. 2002. HHS Responds to Frequently Asked Questions. HIPAAdvisory. Phoenix Health Systems. www.hipaadvisory.com.
10. Gray, Andrew 2002. Risk Evaluation and Management Explained. Information Security Research Notes. North Brunswick, NJ: Icons, Inc. www.iconsinc.com
11. Greenberg, Jerald & Baron, Robert A. 2000. Behavior in Organizations. Seventh Edition. New Jersey: Prentice-Hall.
12. Gue, D'arcy Guerin. 2002. The HIPAA Security Rule: Overview. HIPAAdvisory. Phoenix Health Systems. www.hipaadvisory.com.

13. Katz, Ralph (editor) 1997. The Human Side of Managing Technical Innovation: A Collection of Readings. Oxford: Oxford University Press.
14. Krutz, Ronald L. & Vines, Russell Dean. 2001. The CISSP Prep Guide – Mastering the Ten Domains of Computer Security. New York: Wiley.
15. Mandia, Kevin & Prosise, Chris 2001. Incident Response – Investigating Computer Crime. New York: Osborne/McGraw-Hill.
16. Morabito, J., Sack, I., & Bhate, A. 1999. Organization Modeling – Innovative Architectures for the 21st Century. New Jersey: Prentice-Hall.
17. Murray, W. 2000. Enterprise Security Architecture. Information Security Management Handbook. Fourth Edition. Boca Raton, Florida: CRC Press LLC.
18. Nonaka & Takeuchi. 1995. The Knowledge Creating Company. Oxford: Oxford University Press.
19. Peltier, Thomas R. 2001. Information Security Risk Analysis. Boca Raton, Florida: CRC Press LLC.
20. Poore, Ralph Spencer 2000. Information Law. Information Security Management Handbook Fourth Edition. Boca Raton, Florida: CRC Press LLC. P. 676.
21. Power, Richard 2002. CSI/FBI Computer Crime and Security Survey. Computer Security Issues and Trends Vol. 8 No. 1. San Francisco: Computer Security Institute.
22. Robbins, Stephen P. 1990. Organization Theory – Structure, Design, and Applications. Third Edition. New Jersey: Prentice-Hall.
23. Schein, Edgar H. 1992. Organizational Culture and Leadership. San Francisco: Jossey-Bass Inc.
24. Schneier, Bruce 2000. Secrets & Lies – Digital Security in a Networked World. New York: Wiley.
25. Schultze, Quentin J. 2002. Habits of the High-Tech Heart. Grand Rapids, MI: Baker Academic.
26. Shafritz, Jay M. & Ott, J. Steven 1996. Classics of Organization Theory. Fourth Edition. Orlando, Florida: Harcourt, Brace and Company.
27. Simons, Robert & Davila, Antonio. 1998. How High is Your Return on Management? Harvard Business Review on Measuring Corporate Performance. Massachusetts: Harvard Business School Press. pp 73-97.
28. Tipton, Harold F. & Krause, Micki (editors) 2000. Information Security Management Handbook. Fourth Edition. Boca Raton, Florida: CRC Press LLC.

29. Treese, G. Winfield & Stewart, Lawrence C., 1998. Designing Systems for Internet Commerce. Reading, Mass.: Addison-Wesley.
30. Tudor, Jan Killmeyer 2001. Information Security Architecture – An Integrated Approach to Security in the Organization. Boca Raton, Florida: CRC Press LLC.

**About the Author:**

As Chief Operating Officer, Paul Rohmeyer is directly responsible for managing Icons Inc.'s internal and external operations – from developing customer relationships and guiding technical engagements to developing the technical team. Rohmeyer has years of experience in the IT industry. Among other accomplishments, he led the Strategic Business Intelligence Initiative at AXA Financial Services and was Director of IT Architecture Planning for Bellcore (now Telcordia.) He holds a BA in Economics from Rutgers University, an MBA in Finance from St. Joseph's University, an MS in Information Management from Stevens Institute of Technology and is currently completing a PhD Stevens Institute of Technology, focusing on information security.

# ABOUT THE EDITORS

## SUMIT GHOSH

Sumit Ghosh is the Hattrick Chair Professor of Information Systems Engineering in the Department of Electrical and Computer Engineering at Stevens Institute of Technology in Hoboken, New Jersey. He received his B.Tech. degree from the Indian Institute of Technology at Kanpur, India, and his M.S. and Ph.D. degrees from Stanford University, California. He is the primary author of five original research monographs and has written 90+ transactions/journal papers and 90 refereed conference papers.

## MANU MALEK

Manu Malek is Industry Professor of Computer Science and Director of the Certificate in CyberSecurity Program at Stevens Institute of Technology. Prior to joining Stevens, he was a Distinguished Member of Technical Staff at Lucent Technologies Bell Laboratories. He has held various academic positions in the US and overseas, as well as technical management positions with Bellcore (now Telcordia Technologies) and AT&T Bell Laboratories. He is a fellow of the IEEE, an IEEE Communications Society Distinguished Lecturer, and the founder and Editor-in-Chief of *Journal of Network and Systems Management*. He earned his Ph.D. in EE/CS from University of California, Berkeley. He is the author, coauthor, or editor of seven books and has published more than 50 papers in the areas of computer communications, network design, operations, and management.

## EDWARD A. STOHR

Edward Stohr holds a Bachelor of Civil Engineering degree from Melbourne University, Australia, and M.B.A. and Ph.D. degrees in Information Science from the University of California, Berkeley. He is currently Associate Dean for Research and Academics at the Howe School of Technology Management, Stevens Institute of Technology, Hoboken, New Jersey. His research interests are centered on the problems of developing computer systems to support work and decision-making in organizations. He is the editor of two books on decision support systems and has published articles in many leading journals. In 1992, Professor Stohr served as chairman of the executive board of the International Conference on Information Systems (ICIS). He is on the editorial boards of a number of leading journals in the information systems field and has also acted as consultant to a number of major corporations.

# INDEX